Jianbin Xue
Integration of CAD/CAPP/CAM

Also of interest

Computer-Aided Verification of Coordinating Processes
R.P. Kurshan, 2014
e-ISBN (PDF) 978-1-4008-6404-1

Basics CAD
J. Krebs, 2007
ISBN 978-3-7643-8109-7, e-ISBN (PDF) 978-3-0356-1274-5,
e-ISBN (EPUB) 978-3-0356-1213-4

Digital Processes
M. Hauschild, R. Karzel, /Herausgeber, 2011
ISBN 978-3-0346-0725-4, e-ISBN (PDF) 978-3-0346-1435-1

Journal of Applied Computer Science Methods
2 Issues/Year
ISSN 2391-8241

Jianbin Xue

Integration of CAD/CAPP/CAM

DE GRUYTER

Science Press Beijing

Author
Prof. Jianbin Xue
Nanjing University of Aeronautics and Astronautics
College of Mechanical and Electrical Engineering
No. 29 Yudao Street
210016 Nanjing
China
meejbxue@nuaa.edu.cn

ISBN 978-3-11-057308-4
e-ISBN (PDF) 978-3-11-057309-1
e-ISBN (EPUB) 978-3-11-057321-3

Library of Congress Control Number: 2018934814

Bibliographic Information Published by the Deutsche Nationalbibliothek
The Deutsche Nationalbibliothek lists this publication in the Deutsche Nationalbibliografie; detailed bibliographic data are available on the Internet at http://dnb.dnb.de.

Typesetting: Integra Software Services Pvt. Ltd
Printing and binding: CPI Books GmbH, Leck
Cover image: Eskemar/iStock/Getty Images

www.degruyter.com

Preface

Overview

This textbook includes the fundamental concepts of computer-aided design (CAD), computer-aided process planning (CAPP), computer-aided manufacturing (CAM), and their integration in a generic framework and also to concurrent engineering (CE). As we all know, with the development of computer science and technology, there are many commercial CAD/CAM software packages available in market today, such as Solid Works, Pro/Engineer, Pro/Creo, CATIA, Master CAM, UG-NX, AutoCAD, and so on. Although the syntax of these systems differs from one another, their semantics remains the same. So we just explain the principles of these systems in a generic and syntax-independent manner, and do not discuss their syntax, so that students can use different systems on the basis of this knowledge. The related mathematical models and concepts are discussed to help students understand the working of these systems.

This textbook has also been designed to meet the demands of both practice-oriented and theoretical courses. Students are required to master the basic knowledge of understanding three-dimensional (3D) modeling and viewing, geometric modeling, product design, process planning, manufacturing, product data management (PDM), product life-cycle management (PLM), and CE. Several widely used CAD/CAM systems are mentioned in the context. Some screenshots using these systems are illustrated in a few chapters. Students are encouraged to practice their learning by operating any of the mentioned software packages.

Audience

Students need a comprehensive and complete source of CAD/CAPP/CAM knowledge in order to become proficient in using any of the CAD/CAM systems. The professional engineers also need to have knowledge about these systems. This textbook offers concentrated knowledge and explains the subject matter in a simple, yet comprehensive and coherent, way. This textbook also helps the users get answers to specific questions they have while using CAD/CAM systems and CE.

The audience of this textbook includes the following:

- The students in mechanical engineering, industrial engineering, manufacturing engineering, and mechatronics engineering
- The instructors for CAD/CAPP/CAM courses
- The professional engineers

https://doi.org/10.1515/9783110573091-201

Organization

This textbook consists of eight chapters. The first chapter is introduction. In this chapter, the principle concepts of CAD, CAPP, and CAM are introduced. The history of development of the systems is described. The concept of integration of CAD/CAPP/CAM is also introduced. The following three chapters describe CAD, CAPP, and CAM, respectively. In Chapter 2, the basic CAD knowledge is provided, with commonly used 3D geometric modeling CAD systems and information inside of the CAD systems. The edge-cutting technology about reverse engineering is also included in this chapter. In Chapter 3, the fundamental concepts of CAPP are presented. The variant and generative CAPP systems are introduced. Some well-known CAPP systems are discussed. In Chapter 4, the computer numerical control (CNC) and CAM are discussed. The new concept of 3D printing is also introduced. After understanding these isolated automation islands, the efforts to integrate these systems are described in Chapter 5. Data exchange methods are also listed. Several neural data formats, such as DXF, IGES, STEP, and STEP-NC are discussed. With so many files generated from the systems, PDM is required to deal with these files. In Chapter 6, the functions of PDM system are analyzed. Some well-known PDM systems are introduced. On the basis of the PDM platform, the integration of CAD/CAPP/CAM has got a new approach. The PDM concept is even extended to product life-cycle management. In Chapter 7, concurrent engineering is described, which is used to enhance the effectiveness of product development. The collaborative design and team work are emphasized in the CE design. In the last chapter, some future developments of CAD/CAPP/CAM are proposed.

This textbook has been organized in such a way that each chapter stands on its own, that is, the chapters need not be taught sequentially. Hence, the chapters can be selected depending on the course's focus, purpose, and philosophy.

Example of course syllabus

This book was initially proposed to provide a textbook for the course of integration of CAD/CAPP/CAM at postgraduate level at Nanjing University of Aeronautics and Astronautics. Here is the course syllabus that has been used for 8 years. This can offer some reference for other courses at other levels or other majors.

Course Code: FE**0530009**
Course Title in English: **Integration of CAD/CAPP/CAM**
College and department: College of Mechanical and Electrical Engineering
Semester: 2
Class hours: 40
Teaching methods: Lecture and experiments
Suitable majors: Mechanical Engineering, Mechatronics Engineering

Assessment instruments: Mini-projects and a final examination
Prerequisites: Computer aided design, numerical control programming

Course Objective and Requirements

This course is developed to systematically deliver the principle knowledge about CAD, CAPP, CAM, PDM, and CE to the students at postgraduate levels. The integration methods are studied by analyzing the exchanged information between these isolated systems. The product data model of STEP is one of the ideal solutions. The STEP model contains the information about the whole product life cycle. PDM provides a platform for 3C (which means CAD, CAPP, and CAM) systems integration. The contents of PDM are developing with the progress of 3C technology. On the basis of the integration of CAD/CAPP/CAM, the concept of CE is introduced, as well as the related theories and applications.

During this course, the students are required to master the basic principles of CAD, CAPP, CAM, and PDM, and be familiar with the commercial software packages. Students must know the integration theories and methods of CAD/CAPP/CAM. After taking this course, the logical thinking ability will be enhanced, and the ability to analyze and solve problems will also be improved. It is helpful for the students for project research in the near future.

Acknowledgments

I would like to express my gratitude to the people who directly or indirectly helped me to write this textbook.

I thank the Press of Science staff for their patience and professional help.

My special thanks to the postgraduate school for their financial aids in publishing this textbook for the JINGPIN course.

Last but not least, I am grateful to my family and friends who supported me with their love and encouragement.

If you have any questions or comments about this textbook, please email me at meejbxue@nuaa.edu.cn.

Jianbin Xue
Nanjing University of Aeronautics and Astronautics
No. 29 Yudao Street (210016) Nanjing, China

https://doi.org/10.1515/9783110573091-202

Contents

1 Introduction

Questions before you read

1. Do you know anything about CAD, CAPP, and CAM?
2. What is the whole product life cycle?
3. How do you plan to carry out product design and manufacture?
4. What kinds of software packages do you need in product design and manufacture?

The goal of this chapter is to help the reader understand the basic concept of integration of CAD/CAPP/CAM, their roles in the whole product life cycle, and the brief history of the isolated systems. The concepts of product data management (PDM) and concurrent engineering (CE) are also briefly introduced in this chapter.

1.1 Product design and manufacture

Do you ever want to design and manufacture a product? During your childhood, you must have been attracted by some amazing products. You want to make it. However, your knowledge was limited at that time. You admired the carpenters who could make a number of products of wood. You also admired the blacksmiths who could make tools of steel. And now there are a large number of factories making many products of different materials every day. Do you have such an intention to find out how these products were made of? Here you will be given a general description about the product life cycle.

Product life cycle

In the field of economics, the product life cycle has been divided into four stages: introduction, growth, maturity and decline. But in this textbook, the product life cycle means the product development life cycle, which is a complex procedure, as illustrated in Figure 1.1. The product begins with a need based on customers' demands. From the voice of customers to market analysis, product design, process planning, product manufacture, sales, use and discard. The first two blocks and the last two blocks are mostly related to markets, sales and services, which can be classified into the management field. The product design, process planning and product manufacture are closely related to mechanical engineering, so these three blocks can be classified into the engineering field. The main contents of this textbook cover these three blocks: product design, process planning and product manufacture. With the development of computer science and technology, computers are used to help engineers in different stages of product development. So computer-aided design (CAD) is the automation of product design; computer-aided process planning (CAPP) is the automation of process planning; and computer-aided manufacturing (CAM) is the automation of product manufacture. This textbook introduces the integration of CAD/CAPP/CAM. The interfaces between the product design, process planning and product manufacture are described and analyzed.

https://doi.org/10.1515/9783110573091-001

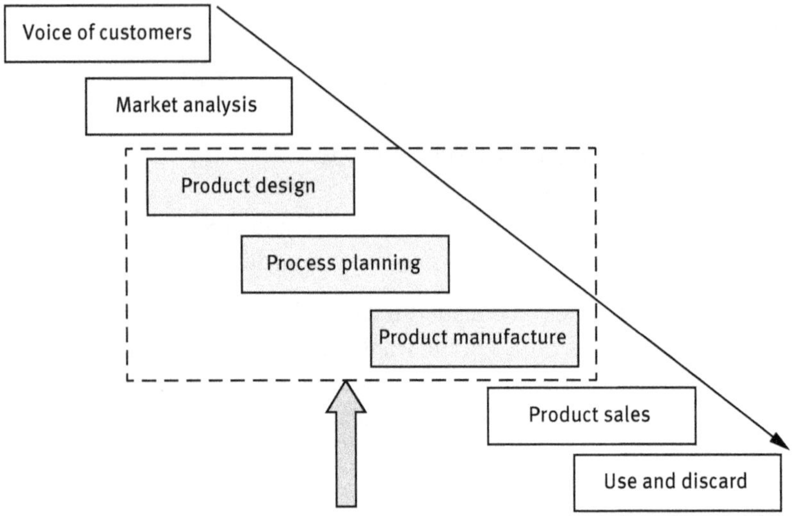

Figure 1.1: Product life cycle.

Basic concept of CAD/CAPP/CAM

Global competitions are increasing in modern manufacturing environment. Products should be delivered with increasing variety, smaller lots and higher quality. Industrial companies cannot survive worldwide competition unless they introduce new products with better quality, at lower costs, and with shorter lead-time. The availability of skilled labor is decreasing. With dramatic changes in computing power and software tools for design and production, engineers are now using CAD, CAPP, CAM, and computer-aided engineering (CAE) systems to automate their design and production processes. And industrial robots are widely used in most manufacturing companies to replace skilled human labor. These technologies are now used every day for engineering tasks. Below is a brief description that how CAD, CAPP, CAM, and CAE technologies are used during the product realization process.

The term CAD/CAPP/CAM is a shortened form of integration of CAD, CAPP and CAM. CAD and CAM are two essential tools to design and manufacture parts. CAPP is trying to bridge the two systems by seamless integration. In recent years, CAE is becoming popular. And it must be also integrated with other CAD/CAPP/CAM systems. So the integration of CAD/CAPP/CAM should be extended to be CAD/CAPP/CAM/CAE. Even more, the integration of CAD/CAPP/CAM should also consider the aspect of robots.

There are several main tasks to be finished, respectively, in each stage. Figure 1.2 shows the main tasks of the three stages.

In the stage of product design, CAD software packages are the main tools. CAD systems are used for geometric modeling, engineering analysis, simulation, scientific computing, graphics, and engineering database. Reverse engineering is a new

Product Design CAD	Process planning CAPP	Product Manufacture CAM
➢ Geometric modeling ➢ Engineering analysis ➢ Simulation ➢ Scientific computing ➢ Graphics ➢ Engineering database ➢ Data communication ➢ Reverse engineering	➢ Rough cast design ➢ Machining process selection ➢ Operation design ➢ Routing design ➢ Ratio of man hour ➢ Tooling ➢ Fixture and jigs	➢ NC programming ➢ Tool path planning ➢ Cutting data file ➢ Simulation of tool path ➢ NC code verification ➢ Check out and trial manufacture ➢ 3D printing

Figure 1.2: Design, process planning and manufacture.

approach for product design. With the development of network, CAD is also used to communicate among or between different engineers from different departments.

In the stage of process planning, CAPP software packages are the main tools. CAPP systems help engineers carry out rough casting design, machining process selection, operation design, routing design, ration of man-hour, tooling, and fixture and jigs.

In the product manufacture stage, CAM software packages are the main tools. CAM systems are used for numerical control (NC) programming, tool path planning, cutting data files, simulation of tool path, NC code verification, check out and trial manufacture. Three-dimensional printing is a new method for product manufacture; it is a kind of additive manufacturing.

Computer applications have been found in the entire spectrum of the product development process, ranging from conceptual design to product realization, and even recycling. CAD, CAPP, and CAM could have been independent systems. They are now seamlessly integrated as CAD/CAPP/CAM because most of the common information about products must be shared among them. So the main point is the integration, which is expressed using the forward slash (/) symbol. Here, the forward slashes among the three or four CAx systems play more important roles in the development of products. They represent the integration relationships among these systems.

1.2 Development of CAD/CAPP/CAM

1.2.1 Computer-aided design

CAD is mainly used for detailed two-dimensional engineering drawings of physical components, and for 3D modeling since the 1980s, but it is also used throughout the engineering process from conceptual design and layout of products, through strength and dynamic analysis of assemblies to definition of manufacturing methods of components.

With the advent of computers, designers have long used computers for their calculations. Initial developments were carried out in the 1960s within the aircraft and automotive industries in the area of 3D surface construction and NC programming, most of it independent of one another and often not publicly published until much later.

First commercial applications of CAD were in large companies such as the automotive and aerospace industries, as well as in electronic industries. Because computers were very expensive, only large corporations could afford the computers capable of performing the calculations.

As computers became more affordable, the application areas have gradually expanded. The personal computers dropped the prices of computers. The development of CAD software for personal computers was the right impetus for almost universal application in all areas.

CAD implementations have evolved dramatically since then. Initially, with 2D in the 1970s, it was typically limited to producing drawings similar to hand-drafted drawings. Advances in programming and computer hardware, notably solid modeling in the 1980s, have allowed more versatile applications of computers in design activities. Some key products for 1981 were the solid modeling packages – Romulus (ShapeData), Uni-Solid (Unigraphics) based on PADL-2, and the release of the surface modeler CATIA (Dassault Systèmes).

Autodesk was found in 1982, which developed a 2D system named Auto CAD. Following the idea of 2D drawing, the computer aided x alliance (CAXA) system was developed to throw off the drawing board in China. The next milestone was the release of Pro/Engineer in 1988, which used the feature-based modeling methods. Also of importance to the development of CAD was the development of the B-rep solid modeling kernels (engines for manipulating geometrically and topologically consistent 3D objects), Parasolid (Shape Data), and ACIS (Spatial Technology Inc.) at the end of the 1980s and the beginning of the 1990s. This led to the release of mid-range packages such as SolidWorks in 1995, Solid Edge (Intergraph) in 1996, and Iron CAD in 1998. Since the beginning of the 21st century, these packages have been enriched with many functional modules. Some big buyouts happened to reorganize the CAD market. In 1997, Dassault Systèmes, best known for its CATIA CAD software, acquired SolidWorks for $310 million in stock. In 2000, Unigraphics purchased structural dynamics research corporation (SDRC) I-DEAS and integrated aspects of both software packages into a single product when became Unigraphics NX. Since 2007, NX has been owned by Siemens PLM Software.

Starting the late 1980s, CAD programs could be run on personal computers. The drafting departments in many small to middle enterprises were massively downsizing. As a general rule, one CAD operator could readily replace at least four or five drafters using traditional ruler–pencil methods. Additionally, many engineers began to do their own drafting work, further eliminating the need for traditional drafting departments. This trend mirrored that of the elimination of many office jobs traditionally performed by a secretary as word processors, spreadsheets, databases, and so on became standard software packages that everyone was expected to learn.

Today, CAD is not limited to drafting and rendering, and it ventures into many more "intellectual" areas of a designer's expertise. CAD is used in many businesses and organizations around the world.

1.2.2 Computer-aided manufacturing

CAD has steadily advanced over the past seven decades to the stage at which designs for new products can be made entirely within the framework of CAD software from the development of the basic design to the bill of materials necessary to manufacture the product. But at this stage, manufacturing has not been considered carefully.

CAM takes this one step further from the conceptual design to the manufacturing of the finished product. Whereas in the past it would be necessary for design developed using 2D CAD software to be manually converted into a drafted paper-drawing detailing instructions for its manufacture, CAM software allows data from 2D CAD software to be converted directly into a set of manufacturing instructions.

Afterward, CAM software converts 3D models generated in CAD into a set of basic operating instructions written in G-code directly. G-code is a programming language that can be understood by numerically controlled machine tools. The G-code can instruct the machine tool to manufacture a large number of items with perfect precision and faith to the CAD design. G-code has a history about 70 years. Figure 1.3 shows a CNC machine tool, which can understand G-code from CAM.

Modern numerically controlled machine tools can be linked into a flexible manufacturing cell (FMC), a collection of tools that each performs a specified task in the manufacture of a product. The product is passed along the cell in the manner of a

Figure 1.3: A CNC machine tool understanding G-code from CAM.

production line, with each machine tool, that is, welding and milling machines, drills, and lathes, performing a single step of the process.

For the sake of convenience, a single computer can drive all of the tools in a single cell. G-code instructions can be fed to this controller and then left to run the cell with minimal input from human supervisors. This may be called as direct numerical control (DNC). If a center computer can drive all of the tools in several cells, the G-code instructions can be delivered to the distributed controllers, this may be called as distributed numerical control (DNC). So DNC has two meanings: one is direct numerical control and another is distributed numerical control.

Benefits of computer-aided manufacturing

The ideal state of affairs for manufacturers is an entirely automated manufacturing process. In conjunction with CAD, CAM enables manufacturers to reduce the costs of producing goods by minimizing the involvement of human operators.

In addition to lower running costs, there are several additional benefits in using CAM software. By removing the need to translate CAD models into manufacturing instructions through paper drafts, it enables manufactures to make quick alterations to the product design, feeding updated instructions to the machine tools and seeing instant results.

In addition, many CAM software packages have the ability to manage simple tasks such as the reordering of parts, further minimizing human involvement. Though all CNC machine tools have the ability to sense errors and automatically shut down, many can actually send a message to their human operators through mobile phones or e-mail, informing them of the problem and awaiting further instructions.

All in all, CAM software represents a trend to make manufacturing entirely automated. While CAD removes the need to retain a team of drafters to design new products, CAM removes the need for skilled and unskilled factory workers. All of these developments result in lower operational costs, lower-end product prices, and increased profits for manufacturers.

Problems with computer-aided manufacturing

Unfortunately, there are several limitations of CAM. Obviously, setting up the infrastructure to begin with can be extremely expensive. CAM requires not only the numerically controlled machine tools themselves but also an extensive suite of CAD/CAM software and hardware to develop the design models and convert them into manufacturing instructions – as well as trained operators to run them.

Additionally, the field of CAM is fraught with inconsistency. While all numerically controlled machine tools operate using G-code, there is no universally used standard for the code itself. Since there is such a wide variety of machine tools that use the G-code, it tends to be the case that manufacturers create their own codes to operate their own machinery.

While this lack of standardization may not be a problem in itself, it can become a problem when the time comes to convert 3D CAD designs into G-codes. CAD systems tend to store data in their own proprietary format, so it can often be a challenge to transfer data from CAD to CAM software and then into whatever form of G-code the manufacturer employs.

Well before the development of CAD, the manufacturing world adopted tools controlled by numbers and letters to fill the need for manufacturing complex shapes in an accurate and repeatable manner. During the 1950s, these numerically controlled machines used the existing technology of paper tapes with regularly spaced holes punched in them to feed numbers into controller machines that were wired to the motors positioning the work on machine tools. The electro-mechanical nature of the controllers allowed digital technologies to be easily incorporated as they were developed.

By the late 1960s, numerically controlled machining centers were commercially available, incorporating a variety of machining processes and automatic tool changing. Such tools were capable of doing work on multiple surfaces of a work-piece, moving the work-piece to positions programmed in advance and using a variety of tools – all automatically. What is more, the same work could be done over and over again with extraordinary precision and very little additional human input. NC machine tools immediately raised automation of manufacturing to a new level once feedback loops were incorporated. The tool tells the computer where it is, while the computer tells it where it should be.

What finally made NC technology enormously successful was the development of the universal NC programming language called automatically programmed tools (APT). Announced at MIT in 1962, APT allowed programmers to develop postprocessors specific to each type of NC machine tool so that the output from the APT program could be shared among different parties with different manufacturing capabilities.

1.2.3 CAD/CAM software

The development of CAD had little effect on CNC initially due to the different capabilities and file formats used by drawing and machining programs. However, as CAD applications such as SolidWorks and Auto CAD incorporate CAM intelligence, and as CAM applications such as Master CAM adopt sophisticated CAD tools, both designers and manufacturers are now enjoying an increasing variety of capable CAD/CAM software. Most CAD/CAM softwares were developed for product development and the design and manufacturing of components and molds.

Today, most of new machine tools incorporate CNC technologies. These tools are used in every conceivable manufacturing sector. CNC technology is related to computer-integrated manufacturing (CIM), CAPP and other technologies such as group technology (GT), and flexible manufacturing cell (FMC). Flexible manufacturing systems (FMS) and just-in-time (JIT) productions are also made possible by numerically controlled machine tools.

1.2.4 Computer-aided process planning

Unlike CAD and CAM systems which have been used in industry for a long time, CAPP systems are yet to be transferred from research laboratories to industry. Aircraft industries are trying to improve their performance by developing CAPP software. Aircraft parts are specific mainly because of their complex geometry, which generates many accessibility problems, and their thin shells, which generate machining deformation and vibration problems. New CAPP approaches have been specified and developed for aircraft structural parts. Machining features are defined as machining faces. The part is split up into a set of machining faces with accessibility, adjacency and rigidity properties. New set-up strategies are defined. These highlight three particular features in the machining of such parts: the categorization, which is a strict guide as to how the part should be machined; the best-fit blank orientation relative to the part; and the potential machining directions of the part, which are the main elements for decision making.

1.3 Product data management

Traditional engineering design created product drawings and schematics on paper and using CAD tools to create part lists. CAM tools generated manufacturing instructions and G-code programming. These engineering data must be managed with certain tools.

PDM manages engineering data – all the data related to a product and to the processes used to design, manufacture and support the product. Much of this data will be created with computer-based systems such as CAD, CAM, and CAE. PDM systems also manage the flow of work through those activities that create or use engineering data. They support techniques, such as concurrent engineering, that aim to improve engineering workflow.

There are now computer systems on the market that help improve the flow, quality and use of engineering information throughout a company. These systems provide improved management of the engineering process through better control of engineering data, of engineering activities, of engineering changes and of product configurations.

1.4 Concurrent engineering

Concurrent engineering is a strategy that replaces the traditional product development process with one in which tasks are carried out in parallel and there is an early consideration for every aspect of a product's development process. This strategy focuses on the optimization and distribution of a firm's resources in the design and development process to ensure an effective and efficient product development process. The major business process should be re-engineered within the organizations and firms that use it, due to the people and process integration requirements. Collaboration is a must for any individuals, groups, departments, and separate organizations within the firm. A firm must be dedicated to the long-term implementation, appraisal, and continuous revision of a concurrent engineering process.

The basics of CAD, CAPP, CAM, PDM, and CE are covered in this textbook. Students should realize that each of the topics deserves a separate book by itself, and they should refer to some existing books for further information.

1.5 Extending integration of CAD/CAPP/CAM

Since the business of product development and manufacturing goes beyond activities of design, process planning and manufacture, integration will not stop at CAD/CAPP/CAM. The integration of CAD/CAPP/CAM system may be extended in various aspects.

Inspection, for example, is a logic step after CNC machining. Closed-loop machining cannot be realized without inspection. The dimensional inspection data model is specified in ISO 10303 AP219.

A CAD system is a combination of hardware and software that enables engineers to design everything from simple part to complex airplanes. It allows an engineer to view a design from any angle with the push of a button and to zoom in or out for close-ups and long-distance views. In addition, the computer keeps track of design dependencies so that when the engineer changes one value, all other values that depend on it are automatically changed accordingly.

CAM system is a type of computer application that helps automate manufacture in a factory. All these systems are concerned with automatically directing the manufacture and inventory of parts.

A frequently overlooked step in the integration of CAD/CAM is the process planning that must occur. CAD systems generate graphically oriented data and may go so far as graphically identifying metal to be removed during processing. In order to produce such things as NC instructions for CAM equipment, basic decisions regarding equipment to be used, tooling and operation sequence need to be made. This is the function of CAPP. Without some elements of CAPP, there would not be such a thing as CAD/CAM integration. Thus, CAD/CAM systems that generate tool paths and NC programs include limited CAPP capabilities or imply a certain approach to processing.

CAD systems also provide graphically oriented data to CAPP systems to use to produce assembly drawings. Furthermore, this graphically oriented data can then be provided to manufacturing in the form of hardcopy drawings or work instruction displays. This type of system uses work instruction displays at factory workstations to display process plans graphically and guide employees through assembly step by step. The assembly is shown on the screen and as an employee steps through the assembly process with a foot-switch, the components to be inserted or assembled are shown on the cathode ray tube (CRT) graphically along with text instructions and warnings for each step.

If NC machining processes are involved, CAPP software exists which will select tools, feeds, and speeds, and will prepare NC programs.

CAPP is a highly effective technology for discrete manufacturers with a significant number of products and process steps. Rapid strides are being made to develop generative planning capabilities and incorporate CAPP into a computer-integrated manufacturing architecture. The first step is the implementation of GT or feature technology (FT) classification and coding. Commercially available software tools currently exist to support both GT and CAPP. As a result, many companies can achieve the benefits of GT and CAPP with minimal cost and risk. Effective use of these tools can improve a manufacturer's competitive advantage.

Understanding the operation of a comprehensive CAD/CAPP/CAM solution requires some study of traditional design and manufacturing practice, a look at the current state of art, and consideration of how technology may change in the future.

Below are some of the commercial packages in the current market.

Auto CAD and Mechanical Desktop are some low-end CAD software systems, which are mainly used for 2D modeling and drawing.

NX, Pro/Engineer, CATIA and I-DEAS are high-end modeling and designing software systems that are costlier but more powerful. These software systems also have CAM and engineering analysis capabilities.

Ansys, Abaqus, Nastran, Fluent, and CFX are packages mainly used for the analysis of structures and fluids. Different software systems are used for different purposes. For example, Fluent is used for fluids, and Ansys is used for structures.

CollabCAD and Alibre design are some of the latest CAD systems that focus on collaborative design, enabling multiple users of the software to collaborate on CAD over the Internet.

You are advised to visit the following websites.

http://www.autodesk.com/

AutoCAD is a 2D and 3D product design software from Auto-desk company. Product design suite is a comprehensive solution delivering 2D and 3D product designs, simulation, collaboration and visualization tools to complete your entire engineering process. The digital prototyping capabilities of the suite can help you design better products, reduce development costs, and get to market faster.

PTC

http://www.ptc.com/

PTC Creo (Previously named as Pro/Engineer) is a scalable, interoperable suite of product design software. It helps teams to create, analyze, view product design and leverage this information downstream utilizing 2D CAD, 3D CAD, parametric, and direct modeling.

http://www.plm.automation.siemens.com/

The NX software integrates knowledge-based principles, industrial design, geometric modeling, advanced analysis, graphic simulation, and concurrent engineering. The software has powerful hybrid modeling capabilities by integrating constraint-based feature modeling and explicit geometric modeling. In addition to modeling standard geometry parts, it allows the user to design complex free-form shapes such as airfoils and manifolds. It also merges solid and surface modeling techniques into one powerful tool set.

http://www.3ds.com/

CATIA is an integrated suite of CAD, CAE and CAM applications for digital product, product life cycle management (PLM) and 3D, which was developed by the French company, Dassault Systèmes. It is considered as the world leader CAD/CAM/CAE software. It was widely adopted by the aerospace, automotive, shipbuilding, and other industries.

1.6 Summary

In this chapter, the product life cycle has been introduced. The roles of CAD, CAPP, and CAM have been described. The integration of CAD/CAPP/CAM is very important for the product design and manufacturing. The PDM system is used to manage the documents of design and manufacture, and to arrange the whole work flow process. Concurrent engineering is a business strategy, which focuses on the optimization and distribution of a firm's resources in the design and development process to ensure an effective and efficient product development process. This chapter is the very beginning to study the integration of CAD/CAPP/CAM.

Questions

1. Please describe the product life cycle by taking a common product as an example.
2. What is the difference between design and drafting?
3. What can computers help a designer in product design?
4. Can you list the names of some commercial CAD/CAM software packages?
5. Please describe the importance of CAPP systems.
6. How can you automate the process of manufacturing?
7. What do you think about PDM?
8. Do you know the concept of concurrent engineering?

Tasks

1. Self-introduction: your name, your supervisor and your majors.
2. Give the reason why you choose this course – the objectives, pre-studied courses and study methods.
3. Make some suggestion to teaching methods.

Further reading material

The Sketchpad system

It is argued that a turning point was the development of Sketchpad system in MIT in 1963 by Ivan Sutherland (who later created a graphics technology company with Dr David Evans). The distinctive feature of Sketchpad was that it allowed the designer to interact with computer graphically: the design can be fed into the computer by drawing on a CRT monitor with a light pen. Effectively, it was a prototype of graphical user interface, an indispensable feature of modern CAD.

Find some stories about the early contributors for CAD.

Some mathematical description work on curves:

In the early 1940s

Isaac Jacob Schoenberg, Apalatequi (Douglas Aircraft)
Roy Liming (North American Aircraft)

In the 1960s and 1970s

The most important work on polynomial curves and sculptured surface
Pierre Bezier (Renault)

Paul de Casteljau (Citroen)
Steven Anson Coons (MIT, Ford)
James Ferguson (Boeing)
Carl de Boor (GM)
Birkhoff (GM) and Garabedian (GM).

References

Dai Tong, Tutorial of CAD/CAPP/CAM. Beijing: The Machinary Industry Press, 1996. (only available in Chinese)

Charles H., International Business Competing in the Global Marketplace, 6th edition. New York: McGraw-Hill, 2007.

Rehg J.A., Kraebber H.W., Computer Integrated Manufacturing, 3rd edition. New York: Pearson Education Ltd., 2004.

Zeid I. (ed.), Tong Bingshu (Trans), Master CAD/CAM.Beijing: Tsinghua University Press, 2007. (only available in Chinese)

2 Computer-aided design

Questions before you read

1. What is the difference between design and drafting?
2. What are the general processes of product design?
3. What are the strategies of top-down design?
4. What are the strategies of bottom-up design?
5. What are the principles inside the 3D CAD software?
6. Do you know the models of wireframe, surface, solid, and feature?
7. Do you know some latest development in CAD technology?

The goal of this chapter is to know the general product design process and some detailed knowledge about CAD, which includes the design steps, the hardware and software in a CAD system, the mathematical models for curves and surfaces, and the four 3D modeling approaches. The reverse engineering process is also discussed in this chapter.

2.1 Introduction

Computer-aided design (CAD), at its most basic, is a geometric modeling system used to produce 2D or 3D engineering drawings in a computer. CAD is the use of a computer to aid in the design process. The design process involves identifying a need, generating possible solutions to meet that need, evaluating each solution to determine its merit, and finally developing a detail model, so it can be built. The computers can aid in most steps in this design process using mathematical and graphic processing power of the computer to assist the engineer in creation, modification, analysis, and display of design. CAD is a general term for the powerful tools used for creating, viewing, and analyzing designs on a computer. In the early days of CAD, the software mimicked the manually produced 2D engineering drawings, but as the graphical power of computer increased, 3D CAD has become the norm.

Currently, the primary use of CAD is in the 3D models from which engineering drawings and CNC part programs are produced almost automatically. A virtual model is developed in the CAD process, without the need to fabricate a prototype, so all the testing and analysis can be performed and a part program can be generated. The major goals of CAD for manufacturing are to increase productivity and create some useful data for manufacturing. CAD helps the designer to visualize a design on the computer screen. The designer can make changes and get immediate feedback on the results. Current CAD software also allows for analysis and testing of components before manufacturing the actual part. This is a very important part of modern manufacturing expertise. The increased use of CAD-based engineering analysis tools such

https://doi.org/10.1515/9783110573091-002

as "finite element analysis" (FEA) has increased the quality of product worldwide. The automobile industry can make cars that last longer, run cooler, and provide more power/economy due to the ability of the design engineer to test models in a computer before making the first part.

The general product design process is first introduced in this chapter and then followed by a brief history of CAD development. The components of CAD systems are analyzed, including both hardware and software packages. The mathematical representations of 3D models are introduced, which include wireframe model, surface model, solid model, and feature model.

2.2 General product design

Product design engineers are responsible for product design, analysis, material selection, as well as design and production documentation. Production engineers add production standards for labor, process, and quality to the shared product data from the design engineers. Engineering release is responsible for product change control. World class companies practice quality at the source to move toward a goal of near defect-free operation. In this section, the general process for product design is first described, the philosophy of concurrent engineering is implanted in design process. Then, the top-down and bottom-up assembly design approaches for product design are introduced.

2.2.1 General process for product design

Before any discussion of CAD, it is necessary to understand the design process in general. The general process for product design is characterized by five basic steps:

1. Conceptualization
2. Synthesis
3. Analysis
4. Evaluation
5. Documentation

The conceptualization process defines the problem on the basis of the stated need and voice of customers. The problem is then divided into two categories: typical and atypical. In the synthesis step, detailed design is added to the general solution produced in conceptualization. Here, the design has sufficient detail to determine how it will perform. In the analysis step, the detailed design is tested and the performance data are collected on as many phases of operation as possible. In the evaluation step, the collected performance data are evaluated. If the evaluation of a product indicates that any part of the product does not meet the performance and design specifications, then alternatives to the design are considered. The last step is documentation.

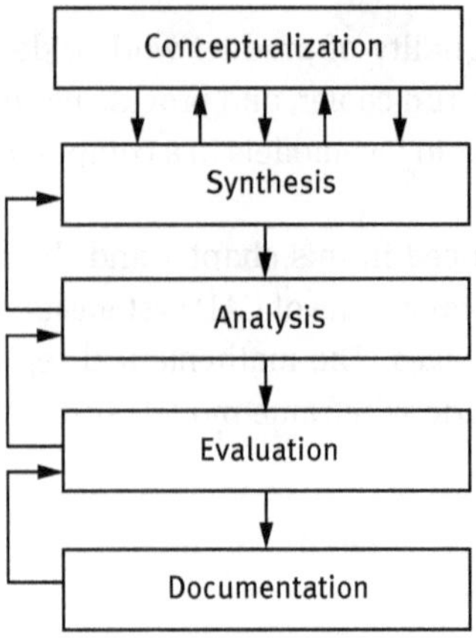

Figure 2.1: The general process for product design.

The final part details are added that permit manufacturing to produce a product that matched design specifications.

The general product design process is illustrated by a sequence of blocks in Figure 2.1. The top three blocks are highly interactive, especially in the early stages of product design. CAD is used at all levels in the design process, with the heaviest use in the design documentation. The software has integrated applications to support the entire design model and works from a shared database of the product.

1. Conceptualization

Conceptualization is the first step of product design. It is ignited by recognition of need of customers and the problems are defined. The characteristics of the product defined in terms of form, fit, and function. The desired shape, style, and character of the product are defined by the term form. The fit describes the relationship of the desired product to other products in the company's line and the degree to which the product matches the target population. The function characteristic defines the product in terms of performance, reliability, maintainability, and other specific order-winning criteria. The form, fit, and function standards for the desired product establish the target limits for the finished design. The data for these three characteristics for new product often come from the marketing area. However, changes to the form, fit, and function for existing product come from a host of internal and external sources, including customer marketing, sales, and design. With the three characteristics identified, the design problem is defined and the design process is started.

A need is usually perceived in one of two ways. Someone must recognize either a problem in an existing design or a customer-driven opportunity in the marketplace for a new product. In either case, the need can be addressed by modifying an existing design or developing an entirely new design. So, the design problem can be defined as either typical or atypical. A typical design problem is one that is similar to previous product design. The atypical design defines a product need that is totally new or different from any previous product. The atypical problems require a concept design approach. The typical problems are handled with repetitive design.

Conceptualization using repetitive design

Repetitive design is the application of the design process to a new product by using most of the previously designed items or small variations from previous designs. Repetitive design process is described by three-step sequence:

(1) Establish all the information data for form, fit, and function.
(2) Categorize current products with similar form, fit, and function characteristics, and then apply parametric analysis to families of parts or assemblies that are similar in form and function by varying in size and detail, or develop a set of standard parts for use across similar products.
(3) Model the product design graphically and analytically to communicate the design configuration effectively.

Parametric analysis and design is a powerful repetitive design tool in CAD programs. It makes the repetitive design easy.

Conceptualization using concept design

Concept design is used to create a new product that is unique, with no similarity to any product currently produced. Conceptual design process depends heavily on the creative nature, or creativity of the designer. The technique of generating a totally unique solution to the present problem is called as thinking out of the box, which requires the inventive, divergent thinking skills. Divergent thinkers think often without the logic seen in convergent thinking, which is usually taught and performed in engineering. There are some useful methods for divergent thinking. Following seven steps are often used for concept design:

(1) Problem statement (Convergent thinking)
(2) Removal of artificial limits and boundaries (Divergent thinking)
(3) Problem definition (C)
(4) Brain-storming (D)
(5) Design selection (C)
(6) Design acceptance (D)
(7) Moving on to synthesis (C)

Here, C indicates that convergent thinking is used predominantly. Convergent thinking, taught in mathematics and science classes, is often called as analytical thinking or deductive reasoning. The solution process moves along a linear path until the first feasiable solution is reached. It is characterized as analytic, judgmental, critical, and selective.

D indicates that divergent thinking is used predominantly. Divergent thinking is characterized as random, non-judgmental, and generative. Divergent thinking moves in many different directions until several solution possibilities are identified. The movement is sometimes linear, but it usually also includes jumping from one solution to another.

Brain-storming can be used to generate multiple ideas and solutions. Some solutions are silly or absurd, while others are logical or practical, but all ideas are accepted as equal at the conceptualization process. This is definitely a divergent thinking process because there are many solutions, not just one.

At the start of every concept design process, the potential for using standard parts, parametric parts, and results from previous designs must be considered along with the requirement for concept design work dictated by a unique form, fit, or functional requirement. Data indicate that pure concept design is required less than 10%. The remaining 90% is divided between repetitive design and some combination of the two design processes.

2. Synthesis

Synthesis is the second step in the design process. The synthesis includes selecting material, adding geometric features, and adding dimensional detail to the design emerging from conceptualization. Synthesis enriches the design, also works as a filter to remove geometric features and material specifications that add cost to the product but not market value using design for assembly and design for manufacturing (DFA&M) methodologies. Conceptualization and synthesis are closely tied and highly iterative. Many up and down arrows represent the information flow between two blocks.

As preliminary product ideas are enriched with features and details early in the design process, the design engineer uses both conceptualization and synthesis skills. As the product design becomes firm, more time is spent in synthesis – adding and verifying product features and details. Almost 70% of the manufacturing cost is determined in conceptualization and synthesis. So, more attention should be paid in these two early stages.

3. Analysis

Analysis is the third step in the design process. It is a method of determining or describing the nature of something by separating it into parts. Analysis involves the study of a single-design solution or several alternative design choices by using mathematical and other scientific procedures. In the analysis process, the elements, or nature of the design, are analyzed to determine the fit between the proposed design and the original design goals.

The types of analysis frequently used fall into three categories: mass properties analysis, tolerance analysis, and finite element analysis (FEA).

Mass properties analysis

Mass properties analysis is often used in the analysis design process. The mass properties analysis frequently used to calculate and return numerical values that describe properties of the selected geometry. In most basic applications, the area of a 2D CAD shape or the volume of a 3D solid object is calculated and displayed. More complex mass properties analysis produces the mass, centroid, moments of inertia, products of inertia, and principal moments with x-y-z directions about centroids. These complex parameters are important for analysis of part geometry that moves and rotates in the final application. Mass properties analysis is important to every type of manufacturer where product design is performed.

Tolerance analysis

Assembly interference, fit, and tolerance analysis is special for analysis of mechanical parts and assemblies design. Based on the existing part geometry, hard gauges are made by manufacturing to check the quality of a manufactured part. With CAE software, soft gauges can be created. For example, if the size and separation distance between two holes is critical for an assembly, a metal gauge like the mating part is produced. This hard gauge is just a metal plate with two pins that can be inserted into the two holes to test the manufactured parts. The soft gauge is a 3D wireframe model of the hard gauge and is created by software as part of the geometry file. When a part is manufactured, the critical features of the finished part are checked electronically by the soft gauge for a proper mating part fit. The soft gauge technique provides better quality monitoring with less investment in hard tooling. Tolerance analysis is used to analyze the fit between mating parts under worst-case tolerance conditions. The tolerance buildup of assembly can be analyzed with an assembly option.

Finite element analysis

Finite element analysis is a numerical program technique for analyzing and studying the functional performance of a structure by dividing the object into a number of small building blocks called finite elements. The FEA process begins with the creation of the geometric model of a part with CAD software. The 3D CAD model is divided into a finite number of small pieces (elements) that are connected to one another at points (nodes). The key to successful FEA is the selection of the mesh, which is composed of elements connected at nodes. The FEA software has mathematical equations that describe how the nodes respond when an external stimulus or force is applied. The size and location of the elements are critical for generation of useful result. FEA software covers a wide range of applications, such as static analysis, transient dynamic analysis, natural frequency analysis, heat transfer analysis, motion analysis, and fluid analysis.

Mass properties analysis, tolerance analysis, and FEA often require that the product or parts in the product should be represented electronically by a solid model. This solid model has all the characteristics of the actual product or part, so the analysis software uses the model to perform analysis.

The line between synthesis and analysis is quite clear. A design developed and then analysis is performed. Analysis is a critical link to ensure that the product meet the criteria required by the customer.

4. Evaluation

Evaluation is the fourth step in product design. It checks the optimum design delivered from the synthesis and analysis process against the original specifications by comparing the form, fit, and function requirements. If the design satisfies the evaluation criteria in every case, the product is passed to documentation. If any areas fall short, the design is returned to the conceptualization, synthesis, or analysis process for more work. Evaluation often requires the construction of a prototype to test for operational performance, reliability, user-friendly operation, and other criteria. Rapid prototyping or 3D printing is a technique to produce a sample product quickly, which is frequently used in this design step.

5. Documentation

Documentation is the last step in design process. Many documents are produced in this stage. All part views are created in the form of working drawings and assembly drawings. The design details, such as standard components, special manufacturing notes, and all dimensions and tolerances, are added in the drawings. The engineering documents consist of part number, bill of material, etc. These electronic data file must be shared by other departments or the electronic data should be stored in PDM database.

This concludes the design model process required to bring a product design from initial concept to complete design. One problem remains with this design model, however. The model process is conducted in isolation because representation from manufacturing, external parts and equipment vendors, and other areas in the enterprise affected by the design are not integrated into the design process and flow. If the design is to be optimal for the initial form, fit, and function requirements, and the manufacturing capability of the enterprise, the design process must be enlarged, and be considered in concurrent engineering environment.

Concurrent engineering implies that the design of a product and the systems to manufacture, service, and ultimately dispose of the product are considered from the initial design concept.

In concurrent engineering environment, participation in the design process is not limited to product design engineers but includes all the specialties listed on the right side of the Figure 2.2. The participants of the concurrent product design include manufacturing engineers, equipment vendors, CNC programmers, fixture designers, quality control, tool designers, even purchasing and marketing engineers, besides product design engineers. The concurrent engineering will be described in detail later in Chapter 8.

2.2.2 Top-down and bottom-up assembly approaches

The product design process can also be classified into top-down and bottom-up approaches according to the assembly design.

The bottom-up approach is most common as it is the traditional and the most logical approach for assembly. In this approach, the individual parts are created

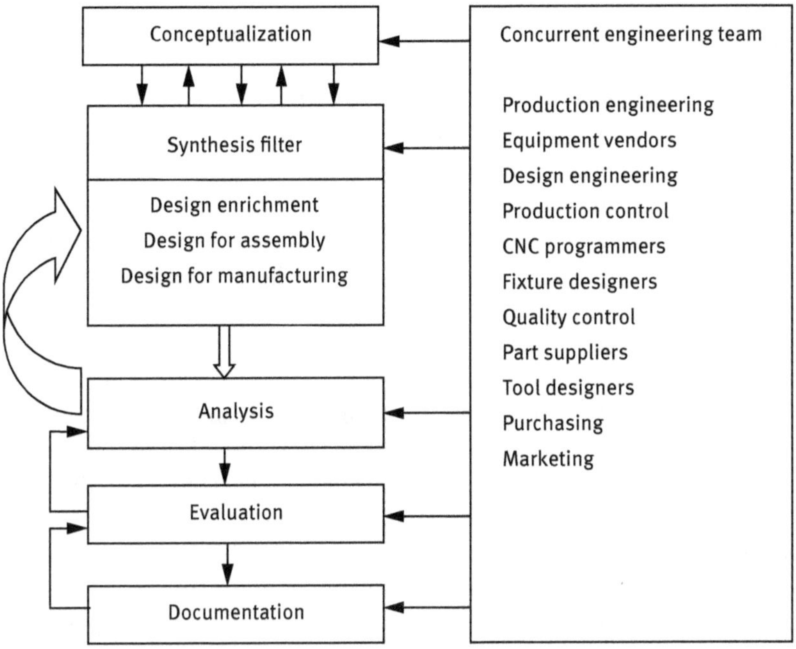

Figure 2.2: Design process in concurrent engineering.

independently, inserted them into an assembly, and using the mating conditions to locate and orient them in the assembly as required by the assembly design.

The assembly modeling process begins with creating a blank assembly model. The parts or subassembly are imported into the model, one at a time. The first part is known as the base part or the host, on top of which other parts are assembled. The proper mating condition must be used to place and orient each inserted part correctly in the assembly mode.

The bottom-up approach has some advantages. It is the preferred technique if the parts have already been constructed. It allows designers to focus on the individual parts. It also makes it easier and simpler to maintain the relationships and regeneration behavior of parts than in the top-down approach. The bottom-up approach appeals to small assemblies.

The top-down approach, while good for any size assembly, is ideal for large assemblies consisting of 10,000 of components. It provides an effective tool and a well-organized approach to manage the design of large complex assemblies. It allows a project leader to break-up product specifications, assign work teams, and enforce downstream design changes at a high level.

The top-down assembly approach becomes a system engineering approach to product design, in which the assembly layout communicates design criteria to subsystem developers, including suppliers. This tight control allows distributed design teams to work concurrently within a common product framework. It also allows detailed design to begin while the assembly layout is being finalized. So, top-down approach fits for concurrent engineering.

The top-down approach plays an important role in the conceptual design phase. It captures the design intent of a product, not only for functions but also for manufacturing, in the early design stages at a high level of abstraction. After all, assembly design does not always require detailed design of constituent parts and subassemblies. This allows the general product designers to validate different design concept before implementing them. The top-down approach also allows designers to practice the what-if design scenarios with ease.

The top-down assembly approach begins with an assembly layout sketch or skeleton model. The layout serves as the backbone of the assembly. The layout defines components in the context of an assembly. These components are empty as they do not have any external reference to actual parts and subassemblies files yet. The assembly layout sketch defines skeletal, space claim, and other physical properties that may be used to define the geometry of and the relationships between components.

The space claim is the most important property of an assembly layout because the layout shows where each assembly component belongs. When a designer lays down the skeletons of all the assembly components in the layout, he can clearly see any interferences, clearances or overlapping between them. The designer can change the location of the component, relative to each other, in the layout to better meet product design requirements.

The top-down approach has many advantages. The major advantage is that if the layout sketch is changed, the assembly and its parts are automatically updated upon exiting the sketch. All the changes are made in one place, that is, the assembly layout sketch.

Either the top-down or the bottom-up approach can be used in product design. The bottom-up approach fits for small assemblies while the top-down approach fits for large complex assemblies. Design engineers decide which approach to use in their product design. And mostly the two approaches can be hybridly used.

2.3 The brief history of CAD development

As we all know, the first computer electronic numerical integrator and calculator (ENIAC) was made in 1946. It was very huge both in terms of size and mass, but it worked slowly. CAD technologies were developed following the development of computer science. The basis of CAD was graphical representation in computers. An early application of computer graphics was used in the SAGE (Semi-Automatic Ground Environment) air defense command and control system in the 1950s. Radar information was converted into computer-generated images on a cathode ray tube (CRT) display. And light pen was used to select information directly from the CRT screen. Is this real CAD? No.

Until 1963, Ivan Sutherland developed Sketchpad as part of his PhD dissertation in MIT. He was considered as the father of CAD. In Sketchpad, the user interacted with the software through a light pen on a large CRT monitor. It was very innovative because computers ran only in batch mode using punched cards and magnetic tapes at that time. Graphical manipulations such as translation, rotation, and scaling could all be accomplished on screen using Sketchpad. But the high cost of computer hardware in the 1960s limited the use of this system to large corporations, such as those in the automotive and aerospace industries, which could justify the initial investment. In those days, CAD systems were just internally developed by manufactures and typically concerned as 2D drafting applications.

With the rapid development of computer technology, computers became more powerful, using faster processor and greater data storage capabilities. Their physical size and cost decreased, and computers became affordable to smaller companies and personal users. In the 1970s, the commercial use of CAD system started. One of the most influential events in the development of CAD was the founding of Manufacturing and Consulting Services Inc. (MCS) in 1971 by Patrick J. Hanratty, who wrote the system Automated Drafting and Machining (ADAM). In 1973, a software company called United Computing purchased the ADAM software code from MCS. The code became a foundation for a product called UNI-GRAPHICS, later sold commercially as Unigraphics in 1975. The Unigraphics System was mostly designed for 2D modeling and drafting. Next year, McDonnell Douglas aircraft bought United Computing. In the same year, Avion Marcel Dassault acquired CADAM from Lockheed and in 1977 they started the development of a 3D CAD named CATI. In 1979, Boeing, General Electric and NIST defined a new 3D data exchange format called initial graphic exchange specification (IGES).

In 1981, Unigrahics introduced their first solid modeling system called UniSolids and Avion Marcel Dassault created their Dassault Systèmes that in the next year was released as CATIA V1 as CADAM add-on. In the same year, I-DEAS was released by SDRC. In 1983, Unigraphics II was introduced to the market. Autodesk released AutoCAD. In 1984, Apple presented the first Macintosh 128, and in the next year, MiniCAD was published, which was the bestselling CAD for Mac. In the middle of 1980s, PCs and Macs were not powerful enough when compared to UNIX. In 1985, Dassault Systèmes released CATIA V2 as a software independent from CADAM. In 1987, Varimetrix produced the first B-rep solid modeling system.

The first release of Pro/Engineer by Parametric Technology Corporation was a big revolution in the CAD industry. It is the first parametric and associative solid modeling system on the market for UNIX workstations. Pro/Engineer first release had also a very innovative and intuitive interface based on X-Window. One year later Unigraphics were also available for UNIX workstation. They retired their UniSolids and released a new program based on Parasolid, that is, UG/Solids. In 1989, the ACIS kernel was also released.

In the early of 1990s, CAD software ran on UNIX workstation or mainframe and minicomputers. In 1991, during a period of financial difficulties, McDonnell Douglas sold Unigraphics to EDS, which at that time was owned by General Motors. The CAD market was dominated by a few companies: IBM-Dassault Systèmes, EDS-Unigraphics, Parametric Technology Corporation and SDRC.

In 1994, Microsoft released their first 32-bit operating system and Intel released their first Pentium Pro. ACIS and Parasolid were quickly available for Windows NT. In 1995, the first SolidWorks released 3D CAD that was available for desktop PC. The advent of new economic Windows-based 3D CAD system heavily modified the market: mid-priced 3D CAD category was born. In 1996, Intergraph released SolidEdge, an ACIS-based CAD, very similar to SolidWorks. Autodesk, whose AutoCAD was losing the market share, released Mechanical Desktop that quickly became the first-selling CAD in the world. In 1997, Dassault Systèmes acquired SolidWorks, and EDS-Unigraphics acquired SolidEdge. In 1998, CATIA V5 was released fully supporting on Windows. In 1999, Autodesk released Inventor, a 3D based on the ACIS kernel and not on AutoCAD (as the previous Mechanical Desktop). In the late 1990s, CAD developers concentrated on improving PDM capabilities and becoming Internet-enabled and no revolutionary technologies appeared.

In 2000, Dassault Systèmes acquired ACIS modeling kernel. In 2001, Unigraphics Solution became UGS and acquired SDRC. In 2002, the new "Next Generation" version of Unigraphics and I-DEAS, called NX, was first released. This will eventually bring the functionality and capabilities of both Unigraphics and I-DEAS together into a single consolidated product. From then on, CAD developers' main efforts were in simplifying and making more intuitive modeling and in integrating CAD in wide PLM suites.

In 2007, SpaceClaim, an innovative history-free direct modeling 3D CAD, was released. Thereafter, many feature-based CAD developers started to integrate direct modeling function in their product. Since 2007, Unigraphics was bought by Siemens, and NX became a part of the Siemens PLM Software. In 2008, NX and SolidEdge integrated a new tool called synchronous technology and SolidWorks proposed Instant 3D. CATIA V6 also released in 2008, allowing direct editing. In 2009, Autodesk launched its Inventor fusion technology.

During the last three years, NX released new versions each year from NX 8 to NX 10.

2.4 Components of CAD systems

CAD can be defined as the use of computer systems to assist in the creation, modification, analysis or optimization of a design. The computer systems consist of the hardware and software to perform the specialized design functions required by the particular user.

Software and hardware are computer-related terms that categorize different types of computer-related paraphernalia. In this section, the hardware and software of CAD systems will be introduced.

2.4.1 Hardware of CAD systems

Just as a draftsman traditionally required pen and ink to bring creativity to appear on the page, there are certain essential components to any working CAD system. As early as in the 1960s, the computers were used for interactive graphics applications. The prohibitively high cost of hardware made general use of interactive computer graphics uneconomical until the 1970s. With the development and subsequent popularity of personal computers, interactive graphics applications now are widespread in homes and workplaces.

CAD systems have become available for many hardware configurations. Most CAD systems have been developed for standard computer systems, ranging from mainframes to microcomputers. Computers vary in size, shape, price, and capabilities. Traditionally, computers are grouped into four categories: supercomputers (such as CRAY-2 from USA and Galaxy series computers from China), mainframes, minicomputers, and microcomputers. Microcomputers, which include the personal computer (PC) and engineering workstations, are desktop size or smaller computers. Microcomputers are quickly becoming more powerful, with greater memory capabilities. A wide range of microcomputers are available with 8- to 128-bit word addressing, several gigabytes of memory, built-in hard disk, CD-ROM, and USB storage. In the 1990s, networks became increasingly common in microcomputer environment. Powerful servers that support massive client–server networks have largely replaced the huge mainframe computers. Even the power of a contemporary PC exceeds that of a mainframe from the 1970s. Furthermore, the computational capability of engineering workstations today exceeds that of most minicomputers. The latest trend is to classify computers as supercomputers, servers, workstations, large PCs and small PCs.

Hardware of CAD systems includes every computer-related object that you can physically touch and handle like keyboard, mouse, disk, screen, printer, wires, central processing unit (CPU), floppy disk, USB ports, and light pen drives. The CAD hardware typically includes the computer, one or more graphics display terminal, plotter, and other peripheral equipment, as shown in Figure 2.3.

Here, the LED monitor, printer, plotter, and 3D printer are output devices, the keyboard, mouse, and scanners are input devices. The USB disk, hard disk, and CD are storage devices. The WiFi/network can be either input or output device depending on the computer getting data from network or store data through network.

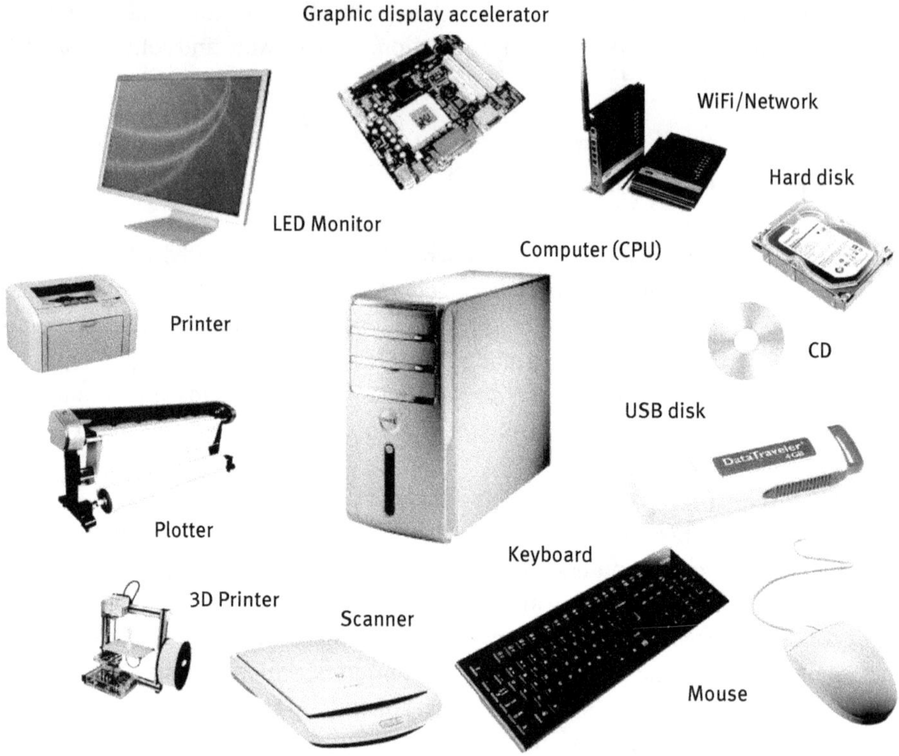

Figure 2.3: Hardware configuration of CAD systems.

2.4.2 Software of CAD systems

Software includes every computer-related program that you cannot feel with the physical senses, for example, system operating system, an anti-virus program, the web browser, all data, documents, and reports. All storage devices that keep data safe and store it in some electronic form are hardware while all data in it is software.

Software is what makes the hardware function properly and to an optimum level. Sometimes, there is confusion between software and hardware because the two terms are so integrally connected. If you buy an anti-virus program you buy software but since it comes on a disk, you have also bought the hardware. The major confusion between software and hardware occurs relating to memory. Software defines the memory capacity of a computer but it depends on the kind of hardware or memory chip used in the particular computer.

Software can be classified into three kinds:

– Applications software: That is of individual liking and need. It can range from small games to professional work-related programs like word processors, and spreadsheets. In CAD domain, the application software include AutoCAD,

Pro/Engineer, CATIA, Solidworks, and ANSYS. They are used to build the graphic models, geometric sculpture, FEA or optimal design.
- Supporting software: In order to manipulate the data or models of CAD, a database is necessary. The database management system is a kind of supporting software. Recently, the retrieve and storage of CAD data files are carried out through network, so the network service systems are also belonged to the supporting software. With the development of cloud computing, some public or private cloud are also considered as the supporting software.
- Systems software: This makes the computer run and, thereafter, makes the applications software on your computer function properly. Without systems software, the applications software cannot be run because the computer needs to be started up with systems software. Systems software is also known as the operating system of a computer. The operating systems include early non-GUI DOS, Unix, and recently Windows NT, Windows XP, and Windows 8 and 10. Another well-known operating system is Linux. Sometimes, the programming language and compiling systems, such as Fortran, Basic, C/C++, Delphi, Java, and Python, are considered as system software. The standard GUIs, such as open Windows in Unix, Windows XP, and Motif X-window (OSF), are considered as system software.

Some computer manufacturers supply the hardware with their own patented systems software. However, some computers can be bought from a provider, and they will run well with systems software bought from another provider. Application software is designed to run on most operating systems. Figure 2.4 shows software configuration of CAD systems.

The CAD software consists of the computer programs to implement computer graphics on the system plus application programs to facilitate the engineering functions of the users.

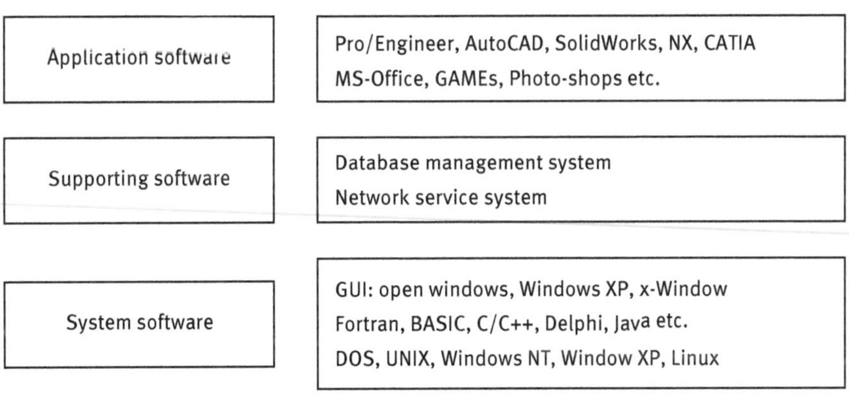

Figure 2.4: Software configuration of CAD systems.

CAD software has applications other than modeling. Examples of these application programs include stress and strain analysis, dynamic response of mechanisms, heat transfer calculations, and numerical control programming.

Modern computers with powerful graphic processors and high-end software make it possible to manipulate almost photo-realistic models with very smooth motion and manipulate the view in sophisticated ways such as pushing a cross-sectional plane through a complex object in real time.

Recently, virtual reality (VR) has become popular. The hardware and software for VR can also be considered as a part of the CAD hardware and software.

2.4.3 Current CAD platform

The PC has been home to CAD software for decades. In the past, CAD software was primarily 2D drafting and 3D wireframe CAD, with a few surface modeling products, such as AutoCAD. While a success story in its own right, AutoCAD, and products like it have never made the step up to providing significant competition to the workstation CAD market. But when the new PCs were developed with Windows NT and fast CPU along with low-priced RAM, workstation CAD users began to envy the low purchase cost, low maintenance cost, ease of use, ease of networking, and performance of the new PCs. Not faster than a workstation yet, but an excellent price versus performance. It did not take long before someone realized these PCs were now capable of running the same solid modeling technology used in the major workstation CAD products. More and more CAD users changed from workstation to PC. Current CAD platforms host on powerful PC. The wireframe, surface, solid, even the feature modeling product can be well performed on PC platform.

2.5 Mathematical models in 3D CAD systems

Geometric modeling is central to all CAD/CAM/CAE applications. Different applications require different representations: some applications use only simple geometric entities such as lines and arcs; others require sculptured curves and surfaces. Here, four kinds of geometric modeling methods are presented.

2.5.1 Wireframe

Wireframe modeling is one of the methods used in geometric modeling systems in CAD. The term “wireframe” comes from designers using metal wire to represent the 3D shape of solid objects. All surfaces are visibly outlined in lines, including the opposite sides and all internal components that are normally hidden from view. Compared

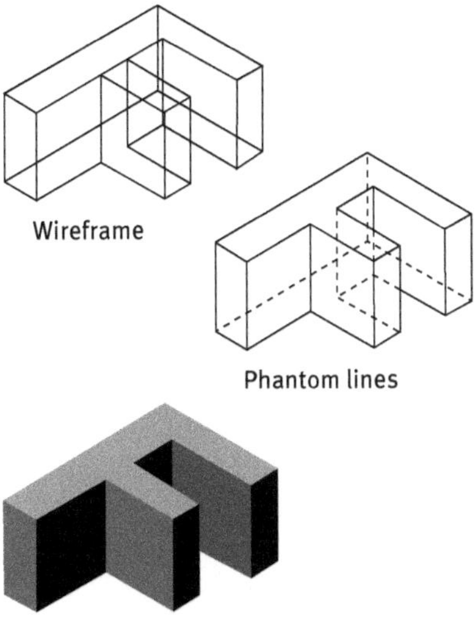

Figure 2.5: A wireframe model.

to surface and solid modeling, wireframe modeling is the least complex method for representing 3D images.

For example in Figure 2.5, the wireframe image at the top shows all hidden lines. In the middle, the hidden lines are turned into dotted lines, known as "phantom lines." The bottom shaded view could be a surface model or a solid model.

A wireframe model represents the shape of a solid object with its characteristic lines and points. Today, wireframe models are used to define complex solid objects. The designer makes a wireframe model of a solid object, and then the CAD operator reconstructs the object, including detailed analysis. The designer can ignore the geometry inside a surface while in solid modeling the designer has to give consistent geometry for all details; wireframe models require less memory space and CPU capacity.

A wireframe model is created by specifying each edge of the physical object where two mathematically continuous smooth surfaces meet, or by connecting an object's constituent vertices using straight lines or curves. The straight lines and arcs are normal curves. Here, four special curves are expressed in matrix mode.

Curves

Hermilton and Furguson curve

$$r(u) = \begin{bmatrix} 1 & u & u^2 & u^3 \end{bmatrix} \begin{bmatrix} 1 & 0 & 0 & 0 \\ 0 & 0 & 1 & 0 \\ -3 & 3 & 2 & -1 \\ 2 & -2 & 1 & 1 \end{bmatrix} \begin{bmatrix} r(0) \\ r(1) \\ r'(0) \\ r'(1) \end{bmatrix} \tag{2.1}$$

where $0 \le u \le 1$.

Bezier curve

$$r(u) = \begin{bmatrix} 1 & u & u^2 & u^3 \end{bmatrix} \begin{bmatrix} 1 & 0 & 0 & 0 \\ -3 & 3 & 0 & 0 \\ 3 & -6 & 3 & 0 \\ -1 & 3 & -3 & 1 \end{bmatrix} \begin{bmatrix} V_0 \\ V_1 \\ V_2 \\ V_3 \end{bmatrix} \tag{2.2}$$

where $0 \le u \le 1$.

Draw a smooth curve through a set of 2D points with Bezier primitives.

B-Spline curve

$$r_i(u) = \begin{bmatrix} 1 & u & u^2 & u^3 \end{bmatrix} \frac{1}{6} \begin{bmatrix} 1 & 4 & 1 & 0 \\ -3 & 0 & 3 & 0 \\ 3 & -6 & 3 & 0 \\ -1 & 3 & -3 & 1 \end{bmatrix} \begin{bmatrix} V_i \\ V_{i+1} \\ V_{i+2} \\ V_{i+3} \end{bmatrix} \tag{2.3}$$

where $0 \le u \le 1$.

NURBS curve

$$R(u) = \sum_{i=0}^{n} V_i^w \cdot N_{i,k}(u)$$

$$r(u) = H\{R(u)\} = \frac{\sum_{i=0}^{n} W_i \cdot V_i \cdot N_{i,k}(u)}{\sum_{i=0}^{n} W_i N_{i,k}(u)} \tag{2.4}$$

where $0 \le u \le 1$.

NURBS is a powerful extension of B-splines, which is a non-uniform rational B-splines.

2.5.2 Surface

Surface model is a mathematical technique for representing solid-appearing objects. Surface modeling is a more complex method for representing objects than wireframe modeling, but not as sophisticated as solid modeling. Surface modeling is widely used in CAD for illustrations and architectural renderings. Figure 2.6 shows coons surface. Here, four special free-form surfaces are expressed in matrix mode.

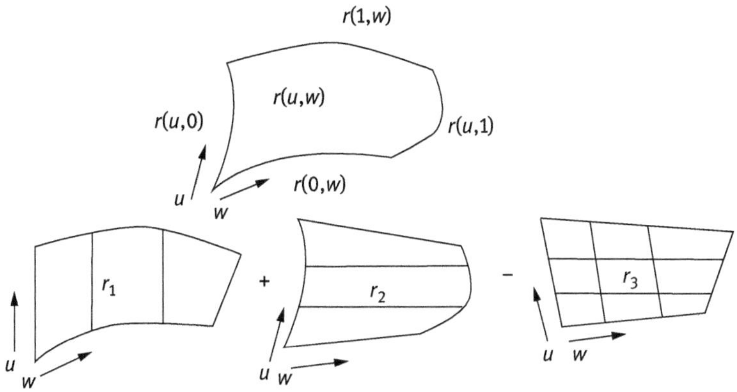

Figure 2.6: Coons surface.

Coons surface

$$r(u,w) = r_1 + r_2 - r_3$$

$$r(u,w) = \begin{bmatrix} -1 & 1-u & u \end{bmatrix} \begin{bmatrix} 0 & r(u,0) & r(u,1) \\ r(0,w) & r(0,w) & r(0,1) \\ r(1,w) & r(1,w) & r(1,1) \end{bmatrix} \begin{bmatrix} -1 \\ 1-w \\ w \end{bmatrix} \tag{2.5}$$

$$r(u,w) = -\begin{bmatrix} -1 & 1-u & u \end{bmatrix} \begin{bmatrix} 0 & r(u,0) & r(u,1) \\ r(0,w) & r(0,0) & r(0,1) \\ r(1,w) & r(1,0) & r(1,1) \end{bmatrix} \begin{bmatrix} -1 \\ 1-w \\ w \end{bmatrix}$$

where $0 \le u \le 1$ and $0 \le w \le 1$.

Bezier surface

$$V = \begin{bmatrix} V_{0,0} & V_{0,1} & V_{0,2} & V_{0,3} \\ V_{1,0} & V_{1,1} & V_{1,2} & V_{1,3} \\ V_{2,0} & V_{2,1} & V_{2,2} & V_{2,3} \\ V_{3,0} & V_{3,1} & V_{3,2} & V_{3,3} \end{bmatrix}$$

$$r(u,w) = \begin{bmatrix} 1 & u & u^2 & u^3 \end{bmatrix} \begin{bmatrix} 1 & 0 & 0 & 0 \\ -3 & 3 & 0 & 0 \\ 3 & -6 & 3 & 0 \\ -1 & 3 & -3 & 1 \end{bmatrix} \begin{bmatrix} V_{0,0} & V_{0,1} & V_{0,2} & V_{0,3} \\ V_{1,0} & V_{1,1} & V_{1,2} & V_{1,3} \\ V_{2,0} & V_{2,1} & V_{2,2} & V_{2,3} \\ V_{3,0} & V_{3,1} & V_{3,2} & V_{3,3} \end{bmatrix} \begin{bmatrix} 1 & -3 & 3 & -1 \\ 0 & 3 & -6 & 3 \\ 0 & 0 & 3 & -3 \\ 0 & 0 & 0 & 1 \end{bmatrix} \begin{bmatrix} 1 \\ w \\ w^2 \\ w^3 \end{bmatrix} \tag{2.6}$$

where $0 \le u \le 1$ and $0 \le w \le 1$.

B-Spline surface

$$V = \begin{bmatrix} V_{0,0} & V_{0,1} & V_{0,2} & V_{0,3} \\ V_{1,0} & V_{1,1} & V_{1,2} & V_{1,3} \\ V_{2,0} & V_{2,1} & V_{2,2} & V_{2,3} \\ V_{3,0} & V_{3,1} & V_{3,2} & V_{3,3} \end{bmatrix}$$

$$r(u,w) = \begin{bmatrix} 1 & u & u^2 & u^3 \end{bmatrix} \frac{1}{6} \begin{bmatrix} 1 & 4 & 1 & 0 \\ -3 & 0 & 3 & 0 \\ 3 & -6 & 3 & 0 \\ -1 & 3 & -3 & 1 \end{bmatrix} \begin{bmatrix} V_{0,0} & V_{0,1} & V_{0,2} & V_{0,3} \\ V_{1,0} & V_{1,1} & V_{1,2} & V_{1,3} \\ V_{2,0} & V_{2,1} & V_{2,2} & V_{2,3} \\ V_{3,0} & V_{3,1} & V_{3,2} & V_{3,3} \end{bmatrix} \begin{bmatrix} 1 & -3 & 3 & -1 \\ 4 & 0 & -6 & 3 \\ 1 & 3 & 3 & -3 \\ 0 & 0 & 0 & 1 \end{bmatrix} \begin{bmatrix} 1 \\ w \\ w^2 \\ w^3 \end{bmatrix}$$

where $0 \le u \le 1$ and $0 \le w \le 1$.

NURBS surface

$$r(u,w) = \frac{\sum_{i=0}^{n_u} \sum_{j=0}^{n_w} W_{i,j} \cdot V_{i,j} \cdot N_{i,k_u}(u)\, N_{j,k_w}(w)}{\sum_{i=0}^{n_u} \sum_{j=0}^{n_w} W_{i,j} \cdot N_{i,k_u}(u)\, N_{j,k_w}(w)}$$

$$r(u,w) = \frac{\sum_{i=0}^{n_u} \sum_{j=0}^{n_w} W_{i,j} \cdot V_{i,j} \cdot N_{i,k_u}(u)\, N_{j,k_w}(w)}{\sum_{i=0}^{n_u} \sum_{j=0}^{n_w} W_{i,j} \cdot N_{i,k_u}(u)\, N_{j,k_w}(w)} \tag{2.8}$$

where $0 \le u \le 1$ and $0 \le w \le 1$.

2.5.3 Solids

Solid modeling techniques allow for the automation of several difficult engineering calculations that are carried out as a part of the design process. Simulation, planning, and verification of processes such as machining and assembly were one of the main *catalysts* for the development of solid modeling. More recently, the range of supported manufacturing applications has been greatly expanded to include sheet metal manufacturing, injection molding, welding, and pipe routing. Beyond traditional manufacturing, solid modeling techniques serve as the foundation for rapid prototyping, digital data archival, and reverse engineering by reconstructing solids from sampled points on physical objects, mechanical analysis using finite elements, motion planning and NC path verification, kinematic and dynamic analysis of mechanisms, and so on. A central problem in all these applications is the ability to effectively represent and manipulate 3D geometry in a fashion that is consistent with the physical behavior of real product. Solid modeling research and development has effectively addressed many of these issues and continues to be a central focus of CAE.

Boundary representation (B-Rep)

In B-rep scheme, a solid is represented with the cellular decomposition of its boundary. Since the boundaries of solids can separate space into regions defined by the interior of the solid and the exterior of the solid, every point in space can be unambiguously tested against the solid by testing the point against the boundary of the solid. Recall that ability to test every point in the solid provides a guarantee of solidity. Using ray casting, it is possible to count the number of intersections of a cast ray against the boundary of the solid. Even number of intersections corresponds to exterior points, and odd number of intersections corresponds to interior points. The assumption of boundaries as manifold cell complexes forces any boundary representation to obey disjointedness of distinct primitives, that is, there are no self-intersections that cause non-manifold points. In particular, the manifoldness condition implies that all pairs of vertices are disjoint, pairs of edges are either disjoint or intersect at one vertex, and pairs of faces are disjoint or intersect at a common edge. Several data structures that are combinatorial maps have been developed to store boundary representations of solids. In addition to planar faces, modern systems provide the ability to store quadrics and NURBS surfaces as a part of the boundary representation. Boundary representations have evolved into a ubiquitous representation scheme of solids in most commercial geometric modelers because of their flexibility in representing solids exhibiting a high level of geometric complexity.

Constructive solid geometric

Constructive solid geometry (CSG) means a family of schemes for representing rigid solids as Boolean constructions or combinations of primitives through the regularized set operations. CSG and boundary representations are currently the most important representation schemes for solids. CSG representations take the form of ordered binary trees where non-terminal nodes represent either rigid transformations or regularized set operations. Terminal nodes are primitive leaves that represent closed regular sets. The semantics of CSG representations is clear. Each sub-tree represents a set resulting from applying the indicated transformations/regularized set operations on the set represented by the primitive leaves of the sub-tree. CSG representations are particularly useful for capturing design intent in the form of features corresponding to material addition or removal (bosses, holes, pockets, etc.). The attractive properties of CSG include conciseness, guaranteed validity of solids, computationally convenient Boolean algebraic properties, and natural control of a solid's shape in terms of high-level parameters defining the solid's primitives and their positions and orientations. The relatively simple data structure and elegant recursive algorithms have further contributed to the popularity of CSG.

Sweeping

Sweeping scheme is very simple. A set moving through space may trace or sweep out volume (a solid) that may be represented by the moving set and its trajectory. A solid is created by extending a profile shape along a specified path. Such a representation is important in the context of applications such as detecting the material removed from a cutter as it moves along a specified trajectory, computing dynamic interference of two solids undergoing relative motion, motion planning. Most commercial CAD systems provide the functionality for constructing swept solids mostly in the form of a 2D cross-section moving on a space trajectory transversal to the section. However, current CAD systems have shown several approximations of 3D shapes moving across one parameter, and even multi-parameter motions. When you sweep a profile along a path, the profile is moved and aligned normal (perpendicular) to the path. You can also specify a value that will change the size of the object from the beginning of the sweep to the end with a mathematical expression to constrain the object scaling. By entering a twist angle, the object can also rotate along the length of the profile with a mathematical expression to constrain the objects twist angle.

2.5.4 Features

Features are defined to be parametric shapes associated with attributes such as length, width, depth, position and orientation, geometric tolerances, material properties, and references to other features. Features also provide access to related production processes and resource models. Thus, features have a semantically higher level than primitive closed regular sets. Features are generally expected to form a basis for linking CAD with downstream manufacturing applications, and also for organizing databases for design data reuse. Parametric feature-based modeling is frequently combined with CSG to fully describe systems of complex objects in engineering. Features such as hole, screw, block, chamfer, and other types of features are widely used in current CAD systems.

A CAD/CAM system may offer different types of design features, such as Protrusion, Slot, Cut, Hole, Round, Chamfer, *Tweak*, and *Cosmetic* features as in Pro/Engineer. Each type of feature exist different kinds of constructing methods. Table 2.1 illustrates these feature types and some of their variations. These features are categorized into five groups: positive, negative, modification, implicit, and patterned features.

2.5.5 An example of 3D modeling

Here, let us take the simple cubic object as an example to illustrate how the CAD model is mathematically established. The cube is made up of eight points, as shown

Table 2.1: Design features.

Feature categories	Design feature	Feature variations
Positive	Protrusion	Extrude
		Revolve
		Blend
		Sweep
Negative	Slots and cuts	
	Holes	Straight – blind
		Straight – through
		Sketched – blind
		Sketched – through
Modification	Rounds	
	Chamfers	
	Tweak features	
Implicit	Cosmetic features	Sketched
		Thread
		Groove
Patterned	Set of any features	

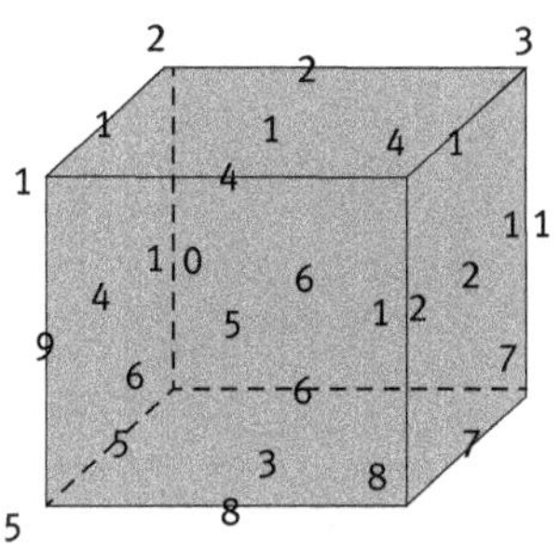

Figure 2.7: a simple cube.

in Figure 2.7. The eight points are listed in Table 2.2. The eight points can form 12 edges of the cube, as shown in Table 2.3. Each of the four edges forms a surface, as shown in Table 2.4.

Only with the eight points, we cannot figure it out what it looks like. With wireframe approach, the shape of a cube has been formed. But the wireframe only shows the shape. We can change the color or thickness of lines. However, there is no surface available, we cannot color the cube and cannot calculate the area of the cube. So if we model the cube with each four lines to construct a surface, then the surface model is constructed. We can now color the surface and calculate the area of the cube. Furthermore, with the surface model, we know nothing about the volume of the cube and the weight of the cube. Is the cube hollow or solid? So we have to further model the cube as a solid. Is the material inside? By now, we have the mass

Table 2.2: List of points.

1	X1	Y1	Z1
2	X2	Y2	Z2
3	X3	Y3	Z3
4	X4	Y4	Z4
5	X5	Y5	Z5
6	X6	Y6	Z6
7	X7	Y7	Z7
8	X8	Y8	Z8

Table 2.3: List of lines.

1	1	2
2	2	3
3	3	4
4	4	1
5	5	6
6	6	7
7	7	8
8	8	5
9	1	5
10	2	6
11	3	7
12	4	8

Table 2.4: List of surfaces.

1	1	2	3	4
2	3	11	7	12
3	5	8	7	6
4	1	9	5	10
5	4	12	8	9
6	2	10	6	11

of the cube if we determine the material of the cube. We have the volume of cube if we determine the dimensions of the cube. These are the mathematical models in CAD systems. And we can even use feature as model to describe the cube, it is the protrude of a square shape.

2.6 Semantics in CAD systems

There are many special terms used in CAD systems. Here, only the commonly used semantics are described to help understand the CAD terminologies.

Coordinate systems

Coordinate systems are the most important concept in a CAD system. They are used to input, store, and display geometric model. There are three kinds of coordinate systems, that is, model coordinate system (MCS), working coordinate system (WCS), and screen coordinate system (SCS).

Sketch planes

A sketch plane is needed to sketch geometry on. Sketch planes are the orthogonal planes created by the axes of MCS or WCS. The sketch planes can be selected by designers.

Layers

Layers are used in digital image editing to separate different elements of an image. A layer can be compared to a transparency on which imaging effects or images are applied and placed over or under an image. In CAD software, a layer is the term used to describe the different levels at which an object or image file is placed. Layers can be partially obscured allowing portions of images within a layer to be hidden or shown in a translucent manner within another image. Layers can also be used to combine two or more images into a single digital image. For the purpose of editing, working with layers allows you to go back and make changes within a layer as you work.

Colors

Color is an attribute of an object, and it can be changed in CAD environment. The color of an object is set either by its layer or by specifying its color explicitly, independent of its layer. Assigning colors by layer makes it easy to identify each layer within the drawing. Assigning colors explicitly provides additional distinctions between objects on the same layer. All objects are created using the current color.

Part modeling

Part modeling is often used to establish a part in digital environment. The designer uses sketch, solid, and features to develop a new part in CAD environment. The part has geometries and topologies. Its information is stored in the CAD database.

Assemble modeling

Assembly modeling is a technology and method used by CAD and product visualization computer software systems to handle multiple files that represent components within a product. The components within an assembly are represented as solid or surface models. The individual data files describing the 3D geometry of individual components are assembled together through a number of sub-assembly levels to create an assembly describing the whole product. All CAD systems support this form of bottom-up construction. Some systems, through associative copying of geometry between components, also allow top-down method of design. Components can be positioned within the product assembly using absolute coordinate placement methods or by means of mating conditions. Mating conditions are definitions of the relative position of components between each other; for example, alignment of axis of two holes or distance of two faces from one another. The final position of all components based on these relationships is calculated using a geometry constraint engine built into the CAD or visualization package.

Engineering drawings

A set of 2D engineering drawings can be generated from 3D models. These drawings are then distributed to the departments and individuals responsible for producing that work. Also these drawings must be with dimensions and tolerances for proper manufacturing. The structure of an engineering drawing includes several projected view and auxiliary view. These views are projected view, sectional view, and detailed view.

Geometric dimensions and tolerances

Geometric dimensions and tolerances are often shortened as GD&T. It refers to a symbolic language used on engineering drawings and computer-generated 3D solid models for explicitly describing nominal geometry and its allowable variation.

The number of terms used in CAD systems is very large. Here, only the commonly used semantics are described to help understand the CAD terminologies. Readers can refer to other technique books for further knowledge about CAD semantics and definitions.

2.7 CAD software packages

There are many CAD software packages available on the markets in these days. Some are available for designers and engineers who are creating a broad variety of products. Others serve niche markets, like those available for bike frame design, jewelry, and home layouts. Here, several well-known CAD software packages will be introduced as follows:

AutoCAD

AutoCAD is a commercial software application developed by Autodesk for 2D and 3D CAD and drafting. AutoCAD was first released in December 1982, running on microcomputers with internal graphic controllers. The native file format of AutoCAD is .dwg. Its interchange file format is .dxf. The DXF file format has become the standard for CAD data interoperability, particularly for 2D drawing exchange.

SolidWorks

SolidWorks is a solid modeling computer program that runs on Microsoft Windows. SolidWorks Corporation was founded in December 1993 by Massachusetts Institute of Technology (MIT) graduate Jon Hirschtick. Hirschtick used $1 million he had made while a member of the MIT Blackjack Team to set up the company. In 1997, SolidWorks was acquired by Dassault for $310 million in stock.

SolidWorks is a solid modeler and utilizes a parametric feature-based approach to create models and assemblies. The software is written based on Parasolid-kernel.

Parameters refer to constraints whose values determine the shape or geometry of the model or assembly. Parameters can be either numeric parameters, such as line lengths or circle diameters, or geometric parameters, such as tangent, parallel, concentric, horizontal, or vertical. Numeric parameters can be associated with each other through the use of relations, which allow them to capture design intent.

Design intent is how the designer of the part wants it to respond to changes and updates. For example, the hole at the top of a beverage can must stay at the top surface, regardless of the height or size of the can. SolidWorks allows the user to specify that the hole is a feature on the top surface and will then honor their design intent no matter what height they later assign to the can.

Features refer to the building blocks of the part. They are the shapes and operations that construct the part. Shape-based features typically begin with a 2D or 3D sketch of shapes such as bosses, holes, and slots. This shape is then extruded or cut to add or remove material from the part. Operation-based features are not sketch-based and include features such as fillets, chamfers, shells, and applying draft to the faces of a part.

Building a model in SolidWorks usually starts with a 2D sketch. The sketch consists of geometry such as points, lines, arcs, conics (except the hyperbola), and splines. Dimensions are added to the sketch to define the size and location of the geometry. Relations are used to define attributes such as tangency, parallelism, perpendicularity, and concentricity. The parametric nature of SolidWorks means that the dimensions and relations drive the geometry, not the other way around. The dimensions in the sketch can be controlled independently, or by relationships to other parameters inside or outside of the sketch.

In an assembly, the analog to sketch relations are mates. Just as sketch relations define conditions such as tangency, parallelism, and concentricity with respect to sketch geometry, assembly mates define equivalent relations with respect to the individual parts or components, allowing the easy construction of assemblies. SolidWorks also includes additional advanced mating features such as gear and cam follower mates, which allow modeled gear assemblies to accurately reproduce the rotational movement of an actual gear train.

Finally, engineering drawings can be created either from parts or assemblies. Views are automatically generated from the solid model and notes; dimensions and tolerances can then be easily added to the drawing as needed.

SolidWorks files use the Microsoft Structured Storage file format. This means that there are various files embedded within each SLDDRW (drawing files), SLDPRT (part files), SLDASM (assembly files) file, including preview bitmaps and metadata sub-files. Various third-party tools can be used to extract these sub-files, although the sub-files in many cases use proprietary binary file formats.

CATIA

CATIA is the acronym of computer-aided 3D interactive application. It is a multi-platform CAD/CAM/CAE software suite developed by the French company Dassault Systèmes. It is written in the C++ programming language. CATIA started as an in-house development in 1977 by French aircraft manufacturer Avions Marcel Dassault. It was later adopted in the aerospace, automotive, shipbuilding, and other industries.

CATIA supports multiple stages of product development, including conceptualization, design, engineering and manufacturing. CATIA facilitates collaborative engineering across disciplines around its 3DEXPERIENCE platform, including surfacing and shape design, electrical fluid and electronics systems design, mechanical engineering and systems engineering.

CATIA enables the creation of 3D parts, from 3D sketches, sheet metal, composites, molded, forged, or tooling parts up to the definition of mechanical assemblies.

PTC Creo or Pro/Engineer

PTC Creo, formerly known as Pro/Engineer, is a 3D CAD/CAM/CAE feature-based, associative solid modeling software. It is one of a suite of 10 collaborative applications that provide solid modeling, assembly modeling, 2D orthographic views, FEA, direct and parametric modeling, sub-divisional and NURBS surfacing, and NC and tooling functionality for mechanical designers.

Pro/Engineer was the first rule-based constraint, sometimes called "parametric," 3D CAD modeling system in industry. The parametric modeling approach uses parameters, dimensions, features, and relationships to capture intended product behavior and create a recipe which enables design automation and the optimization of design and product development processes. This design approach is used by companies whose product strategy is family-based or platform-driven, where a prescriptive design strategy is fundamental to the success of the design process by embedding engineering constraints and relationships to quickly optimize the design, or where the resulting geometry may be complex or based upon equations. Creo Elements/Pro provides a complete set of design, analysis and manufacturing capabilities on one, integral, scalable platform. These required capabilities include Solid Modeling, Surfacing, Rendering, Data Interoperability, Routed Systems Design, Simulation, Tolerance Analysis, and NC and Tooling Design.

Creo Elements/Pro can be used to create a complete 3D digital model of manufactured products. The models consist of 2D and 3D solid model data which can also be used downstream in FEA, rapid prototyping, tooling design and CNC manufacturing. All data are associative and interchangeable between the CAD, CAE, and CAM modules without conversion. A product and its entire bill of materials (BOM) can be modeled accurately with fully associative engineering drawings, and revision control information. The associativity functionality in Creo Elements/Pro enables users to make changes in the design at any time during the product development process and automatically update downstream deliverables. This capability enables concurrent engineering – design, analysis, and manufacturing engineers working in parallel – and streamlines product development processes.

Siemens-NX or Unigraphics UG

NX, formerly known as NX Unigraphics or usually just UG, is an advanced high-end CAD/CAM/CAE software package originally developed by Unigraphics, but in 2007 acquired by Siemens PLM Software. Its key functions include the following:

- Computer-aided design (CAD) (Design)
  - Parametric solid modeling (feature-based and direct modeling)
  - Free-form surface modeling
  - Reverse engineering
  - Styling and computer-aided industrial design
  - Engineering drawing
  - Product and manufacturing information (PMI)
  - Reporting and analytics, verification and validation
  - Knowledge reuse, including knowledge-based engineering
  - Sheet metal design
  - Assembly modeling and digital mockup
  - Routing for electrical wiring and mechanical piping
- Computer-aided engineering (CAE) (Simulation)
  - Stress analysis/finite element method (FEM)
  - Kinematics
  - Computational fluid dynamics (CFD)
  - Thermal analysis
- And Computer aided manufacturing (CAM) (Manufacturing)
  - Numerical control (NC) programming

CAXA

CAXA is the market leader of industrial software and service in China, as well as a pioneer of China Industrial Cloud service, which mainly supplies with 2D and 3D design softwares (CAD), MES, product life cycle management solution (PLM) and industrial cloud service platform.

With skilled developing team, CAXA keeps on creating and innovating softwares of 2D, 3D CAD and PLM platform. As the first domestic CAD software developing company in China, it has three development centers in Beijing, Nanjing, and the United States, owning more than 150 pieces of copyrights and patents, and it takes part in several custom work of national CAD and CAPP Technical Standard.

CAXA has its independent intellectual property rights, and it is now working to develop and market application software and industrial cloud service, covering fields of product design, process, manufacturing and management. CAXA portfolio consists of PLM solution to digital design and PLM solution to digital manufacturing, such as design (2D&3D CAD), process planning (CAPP), data management (PDM), and manufacturing (CAM, DNC MES, and MPM) to realize coordinative management for product life cycle. Industrial cloud service platform mainly includes cloud design, cloud manufacturing, cloud coordination, cloud resource, and cloud storage, and such tools and services are essential to product innovating.

In China, beside CAXA, there are other CAD software systems, such as GaoHua CAD, ZOOM CAD, HaoChen CAD, KM CAD, and JinYinHua CAD. A CAD system named superman was developed in Nanjing University of Aeronautics and Astronautics.

2.8 Reverse engineering

Reverse engineering is a practical method to create 3D virtual models of an existing physical part. It can be used in 3D CAD software. The reverse engineering process involves measuring an object and then reconstructing it as a 3D model. The physical object can be measured using 3D scanning technologies, such as CMMs, laser scanners, structured light digitizers or industrial CT scanning (computed tomography). The measured data alone, usually represented as a point cloud, lacks topological information and is therefore often processed and modeled into a more usable format such as a triangular-faced mesh, a set of NURBS surfaces, or a 3D CAD model.

Reverse engineering is also used by businesses to bring existing physical geometry into digital product development environments, to make a digital 3D record of their own products, or to assess the products of their competitors. It is used to analyze, for instance, how a product works, what it does and what components it consists of, estimate costs and identify potential patent infringement, etc.

The camera, which is installed on an unmanned aerial vehicle (UAV), can take photos for large objects, such as bridges, buildings and tunnels, to generate point cloud. And using reverse engineering, the digital model of current status of the large objects compared with the original one, the deformation of the objects can be calculated. So, the maintenance plans can be scheduled according to the degree of deformation.

2.9 Product data exchange

2.9.1 GKS

Graphical Kernel System (GKS) is the first ISO standard for low-level computer graphics, introduced in 1977. GKS provides a set of drawing features for 2D vector graphics suitable for charting and similar duties. The calls are designed to be portable across different programming languages, graphics devices and hardware, so that applications written to use GKS will be readily portable to many platforms and devices.

GKS was fairly common on computer workstations in the 1980s and early 1990s. It was little used outside these markets and is essentially obsolete today. The functionality of GKS is wrapped up as a data model standard in the STEP standard, section ISO 10303-46. A descendant of GKS was PHIGS.

2.9.2 PHIGS

Programmer's Hierarchical Interactive Graphics System (PHIGS) is an API standard for rendering 3D computer graphics, considered to be the 3D graphics standard for the 1980s through the early 1990s. PHIGS was designed in the 1980s, inheriting many of its ideas from the GKS of the late 1970s, and became a standard by 1989: ANSI (ANSI X3.144-1988), FIPS (FIPS 153) and then ISO (ISO/IEC 9592 and ISO/IEC 9593). Due to its early gestation, the standard supports only the most basic 3D graphics, including basic geometry and meshes.

PHIGS originally lacked the capability to render illuminated scenes and was superseded by PHIGS+. PHIGS+ works in the same manner, but added methods for lighting and filling surfaces within a 3D scene. PHIGS+ also introduced more advanced graphics primitives, such as NURBS surfaces.

2.9.3 Open GL

OpenGL describes an abstract API for drawing 2D and 3D graphics. Although it is possible for the API to be implemented entirely in software, it is designed to be implemented mostly or entirely in hardware. The API is defined as a number of functions which may be called by the client program, alongside a number of named integer constants (e.g., the constant GL_TEXTURE_2D, which corresponds to the decimal number 3553). Although the function definitions are superficially similar to those of the C programming language, they are language-independent. As such, OpenGL has many language bindings, some of the most noteworthy being the JavaScript binding WebGL (API, based on OpenGL ES 2.0, for 3D rendering from within a web browser); the C bindings WGL, GLX and CGL; the C binding provided by iOS; and the Java and C bindings provided by Android.

In addition to being language-independent, OpenGL is also platform-independent.

2.9.4 Direct X

Direct X is a collection of APIs for handling tasks related to multimedia on Microsoft platforms. Originally, the names of these APIs all began with Direct, such as Direct3D,

DirectDraw, DirectMusic, DirectPlay, and DirectSound. The name Direct X was coined as shorthand term for all of these APIs (the X standing in for the particular API names) and soon became the name of the collection. Direct3D is also used by software applications for visualization and graphics tasks such as CAD/CAM engineering.

2.9.5 IGES

The IGES was jointly developed by industry and the national institute for standard and technology (NIST) in the 1970s. The drawing files were usually translated between different software vendors. Due to the complex drawing files, it was difficult to translate without any errors. Many software packages support the IGES standard. In China, IGES became a national standard, GB/T14213-1993, now is updated to GB/T 14213-2008. IGES has been widely used for data exchange between different CAD/ CAM systems. However, the inherent defect of IGES makes it difficult to use in the field of automatic NC programming. While far from perfect, the IGES standard is the best-supported common format for 3D CAD models at that time. When IGES is used to bring part geometry into the CAM software, an IGES translator is required to convert the standard IGES file format into a format compatible with the CAM software.

2.9.6 STEP/PDES

PDES means product data exchange using STEP. It was an IGES research project to develop the next generation CAD/CAM data exchange standard started in 1984. STEP is the acronym of STandard for the Exchange of Product model data, started in 1985 by ISO. Their goal is to develop a complete, unambiguous computer-interpretable definition of the physical and functional characteristics of a product throughout its life cycle. The ISO 10303 STEP standard for product data exchange is the most popular method to implement translations between CAD and CAM systems. More information and knowledge about STEP will be shared in Chapter 5.

2.10 Kernels of 3D CAD systems

2.10.1 Parasolid

Parasolid is a geometric modeling kernel originally developed by ShapeData, now owned by Siemens PLM Software, that can be licensed by other companies for use in their 3D computer graphics software products.

Parasolid has many powerful capabilities including model creation and editing utilities, such as Boolean modeling operators, feature modeling support, advanced surfacing, thickening and hollowing, blending, and filleting and sheet modeling. Parasolid also includes tools for direct model editing, including tapering, offsetting, geometry replacement and removal of feature details with automated regeneration of surrounding data. Parasolid also provides wide-ranging graphical and rendering support, including hidden-line, wireframe and drafting, tessellation and model data inquiries.

When exported from the parent software package, the Parasolid file has an extension of .x_t. Another file format is .x_b, which is binary so it is more machine-independent. Most Parasolid files can communicate and immigrate only 3D solids and/or surface data – Parasolid files currently cannot communicate and migrate 2D data such as lines and arcs. To use Parasolid effectively, users need to have fundamental knowledge of computational geometry and topology.

2.10.2 ACIS

ACIS is a 3D geometric modeling kernel developed by Spatial Corporation. ACIS is used by many software developers in industries such as CAD, CAM, CAE, coordinate-measuring machine (CMM), 3D animation and shipbuilding. ACIS provides software developers and manufacturers the underlying 3D modeling functionality.

ACIS uses an open object-oriented C++ architecture that enables robust, 3D modeling capabilities. ACIS is used to construct applications with hybrid modeling features, since it integrates wireframe, surface and solid modeling functionality with both manifold and non-manifold topology, and a rich set of geometric operations. In late 2000, Spatial Corporation was acquired by Dassault Systèmes , the ACIS file format changed slightly and was no longer openly published.

ACIS saves modeling information to external files which have an open format allowing external applications, even those not based on ACIS, access to the ACIS geometric model. When reading, restoring or operating a ACIS file, the basic information needed to understand includes the structure of the save file format, how data are encapsulated, the types of data written and subtypes and references.

ACIS file types

ACIS supports two kinds of save files: .sat and .sab. Sat is a standard ACIS text and Sab is a Standard ACIS Binary. The two formats store identical information, so the term Sat file is generally used to refer to either when no distinction is needed.

In the narrow sense, Sat files are ASCII text files that may be viewed with a simple text editor. A Sat file contains carriage returns, white space and other formatting that makes it readable to the human eye.

Sab files cannot be viewed with a simple text editor and are meant for compactness and not for human readability. A Sab file uses delimiters between elements and binary tags, without additional formatting.

Structure of the Sat file

A Sat file contains
- a three-line header
- entity records, representing the bulk of the data
- optionally, a begin history data marker
- optionally, old entity records needed for history and rollback
- optionally, an end history data marker
- an end marker

Beginning with ACIS Release 6.3, it is required that the product ID and units to be populated for the file header before you can save a Sat file.

2.11 CAE – computer aided engineering

CAE is the broad usage of computer software to aid in engineering analysis tasks, such as FEA), CFD, Multibody dynamics (MBD), and optimization. They are widely used in many fields, such as automotive, aviation, space, and shipbuilding industries. CAE tools are being used to analyze the robustness and performance of components and assemblies. The term CAE encompasses simulation, validation, and optimization of products and manufacturing tools. In the future, CAE systems will be major providers of information to help and support design teams to make decision.

2.11.1 Finite element analysis

FEA is a computational tool for performing engineering analysis, such as stress analysis on components and assemblies. It includes the use of mesh generation techniques for dividing a complex problem into small elements, as well as the use of software program coded with FEM algorithm. In applying FEA, the complex problem is usually a physical system with the underlying physics such as the beam equation, or the heat equation expressed in integral equations, while the divided small elements of the complex problem represent different areas in the physical system.

2.11.2 Computational fluid dynamics

Computational fluid dynamics (CFD) is a branch of CAE software. It is a simulation software allowing to predict the impact of fluid flows on the product with confidence. It can be used throughout design and manufacturing as well as during end use. The software's unparalleled fluid flow analysis capabilities can be used to design and optimize new equipment and to troubleshoot in some already existing installations. It can deal with many phenomena – single- or multi-phase, isothermal or reacting, compressible or not.

2.11.3 Kinematic and dynamic analysis

Product designers can evaluate and manage the complex interactions between motion, structures, actuation and controls to better optimize product designs for performance, safety and comfort. Traditionally, the studies of motion and structures are divided into two categories: kinematics and dynamics. Kinematics is the study of motion without regard to forces; dynamics is the study of motions that result from forces. Kinematic simulations show the physical positions of all the parts in an assembly with respect to the time as it goes through a cycle. Dynamic simulation is more complex because the problem needs to be further defined for the forces. But dynamics are often required to accurately simulate the actual motion of a mechanical system. Generally, kinematic simulations help evaluate form, while dynamic simulations assists in analyzing function. Utilizing multibody dynamics solution technology, non-linear dynamics can be analyzed in a fraction of the time required by FEA solutions. Loads and forces computed by simulations improve the accuracy of FEA by providing better assessment of how they vary throughout a full range of motion and operating environments.

2.12 Summary

This chapter introduces the general concept of CAD. The general product design procedure is described. The detail components of CAD systems are introduced with their hardware and software. The mathematical models of curves and surfaces are listed in matrix forms. Some commercial CAD software packages are introduced. Product data exchange is discussed with several graphic kernels. The parasolid and ACIS kernels are also introduced. The CAE is considered as part of CAD system, but with different functions.

The information from CAD system can be summarized as following:
- Geometry information
- Dimensions and tolerances
- Special requirements

In principle, CAD could be applied throughout the design process, but in practice, its impact on the early stages, where very imprecise representations such as sketches are used extensively, has been limited. It must also be stressed that at present CAD does little in helping a designer in a more creative and intuitive way such as generation of possible design solution, or in those aspect that involve complex reasoning about the design by visual examination of drawings, whether the component may be made easy or whether it matches the specifications. In practicing concurrent engineering, there is a pressing need for CAD systems to interface or integrate design with all the down-stream activities.

Questions

1. Define CAD.
2. List the basic components associated with a CAD system.
3. Describe the general process of product design.
4. What are top-down and bottom-up approaches for assembly design?
5. List some major vendors supplying CAD software packages.
6. Take a kind of CAD software as an example, describe its main functions.
7. What are wireframe, surface, solid, and feature modeling approaches?
8. What is constructive solid geometry (CSG)?
9. What are the product data exchange approaches?
10. How does computer-aided engineering help the designers in decision-making?

Tasks

1. Retrieve some memory that you designed the gear box for a transmission system, and explain the difference between product design and engineering drawing.
2. Use the CAD system you are familiar with to design a simple product, such as a vase, a cup, or just a washer. Show your design in wireframe, surface, solid, or feature view.

References

An D., Leep H.R., Parsaei H.R., A product data exchange integration structure using PDES/STEP for automated manufacturing applications. Computers Industrial Engineering, 1995, 29(4): 711–715.

Groover M.P., Automation, Production Systems, and Computer Integrated Manufacturing, 4th edition. New York: Pearson Prentice Hall, 2014.

Rehg J.A. Kraebber H.W., Computer Integrated Manufacturing, 3rd edition. New York: Pearson Prentice Hall, 2005.

Xu X., Integrating advanced computer aided design, manufacturing, and numerical control, principles and implementations. Information Science Reference, 2007.

3 Computer-aided process planning

Questions before you read

1. What activities are involved in the general manual process planning?
2. What is the classification of CAPP systems?
3. What is the group technology?
4. What is the feature-based technology?
5. Why do most researchers still focus on process planning?

The goal of this chapter is to understand the concept of process planning, which is the interface or middleware between design and manufacturing. Process planning is used for transforming a raw material into a finished component. The tasks of manual process planning are described in detail. The CAPP systems have been developed to help the automation of process planning and are considered as the bridge between CAD and CAM. The importance of process planning is investigated. The variant CAPP and the generative CAPP systems are introduced. The coding systems, group technology, and feature-based technology are introduced. Some new approaches such as expert system, neutral network are also introduced. The basis of CAPP and the benefits of implementing CAPP systems are also discussed.

3.1 Introduction

When the design process is finished, companies have gotten the digital 3D models, 2D engineering drawings, and a set of documents of the products. Process planning is one of the most important steps in converting a design model into a manufactured product. It is concerned with determining the sequences of individual machining operations needed to produce a given part or product. The essence of that task is the creation of a complete information package on how to perform the manufacturing process, which may include work instructions, a bill of material (BOM), a quality control plan, tooling planning, and so on for the shop floor. The resulting operation sequence is documented on a form, typically referred as a route sheet, which contains a listing of the production operations and associated machine tools for a work part or assembly. Traditionally, process planning tasks are undertaken by manufacturing process experts. These experts use their experience and knowledge to generate instructions for the manufacture of the products based on the design specification and the available installations and operators. Different process planners may come up with different plans when facing the same problem, leading to inconsistency in process planning and manufacturing.

It is widely accepted that process planning is the bridge between design and manufacture and its role is crucial for an effective integration of design and

https://doi.org/10.1515/9783110573091-003

manufacturing. Computer-aided process planning is a key point factor for really automated computer-aided manufacturing (CIM) systems.

Process planning involves the preparation for the manufacture of products. Process planning deals with the selection and definition of the processes that have to be performed in order to transform raw material into a given shape. The process planning activities include interpretation of design data, selection, and sequencing of operations to manufacture the product, selection of machine and cutting tools, determination of cutting parameters, choice of jigs and fixtures, and the calculation of the machining times and costs.

Process planning is defined as the activity of deciding which manufacturing processes and machines should be used to perform the various operations necessary to produce a component and the sequence that the process should follow. It translates design information into the process steps and instructions to efficiently and effectively manufacture products from raw material to finished product. Figure 3.1 shows a diagram of process planning.

Process planning refers to a set of instructions that are used to make a component or a part so that the design specifications are met. It is defined by the Society of Manufacturing Engineering as follows:

Process planning is the systematic determination of the methods by which a product is to be manufactured economically and competitively.

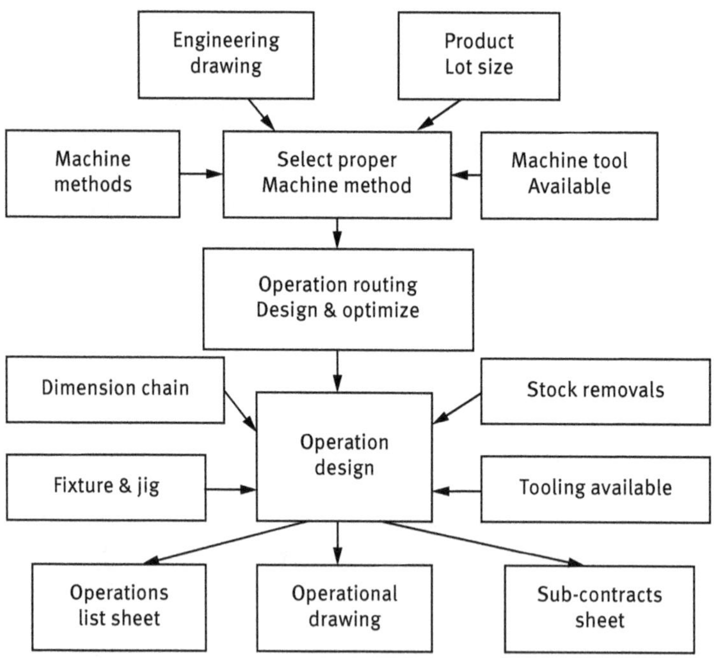

Figure 3.1: The diagram of process planning.

3.2 Manual process planning

Process planning is used to prepare a detailed set of plans and instructions to produce a part from the raw material. The planning begins with 3D models, engineering drawings, specifications, parts or material lists, and a forecast of demand. The process planning must be finished by the deadline.

Process planners use parts lists or BOMs to develop and communicate manufacturing requirements and to indicate how and in what order the product should be built. The development of a process plan involves a number of activities, which include the following:

1. read the engineering drawing, analyze the machining requirements;
2. consult data and handbooks for references;
3. select machine methods and optimize routing plan;
4. choose machine equipment and design fixture;
5. determine machine parameters (cutting speed, feed, and depth of cut);
6. calculate operational dimensions and tolerances;
7. calculate the manufacturing times (set up time, process time, and lead time)
8. draw operational process drawings;
9. fill in process cards and other forms.

The results of the planning are as follows:

- Routings specify operations, operation sequences, work centers, standards, tooling, and fixtures. This routing becomes a major input to the manufacturing resource planning (MRP-II) system to define operations for production activity control purposes and define required resources for capacity requirements planning purposes.
- Process plans typically provide more detailed, step-by-step work instructions including dimensions related to individual operations, machining parameters, set-up instructions, and quality assurance checkpoints.
- Fabrication and assembly drawings to support manufacture (as opposed to engineering drawings to define the part).

Manual process planning is based on a process planner's experience and knowledge of production facilities, equipment, their capabilities, processes, and toolings. Process planning is more time consuming, and the results vary based on the person doing the planning.

Manufacturing routings are developed to decrease lead-time. Components on BOMs, which are attached to a routing operation, identify where specific material should be issued and consumed in the manufacturing sequence of the product. Material planners, who usually drive planning and control within the company, use BOM to determine what manufactured and purchased items are required. Routings are used with the bills of material to determine when, where, and in what quantities parts required, and what resources and work centers are required to complete the work orders.

When a work order is created, the engineering BOM becomes the parts list on the work order. The stockroom uses this parts list to pull parts for the work order. Components are issued for the work order and then back-flushed so that the inventory is relieved.

What is a routing?

A routing lists the operations required to manufacture a product in sequence. Each operation within the routing identifies specific information such as the work center, and time standards for setup, machine run, and labor hours. It may also include additional information such as required tools and inspections. Each part in a BOM can be linked to a routing operation to identify the specific routing operation where a part should be issued from inventory and consumed by the product.

Work center

Work centers are the specific, physical locations on the shop floor where routing operations occur. A work center defines basic information such as the machines and number of people employed at the work center. Additional information can include work center rates for labor, machines, setup, work center capacity, and machine efficiency.

The outcomes of process planning are the method sheets that list the manufacturing instructions for various machine tools to produce the finished part. The information in these sheets can be utilized to program machine tools to produce the part.

3.3 Brief history of CAPP

The study of computer-aided process planning (CAPP) has been getting the research interesting since the later 1960s. The early intent was to establish a computer-aided system, which included the generation of the process card, the storage of the process content, and the search of the process. It uses the computer to store, organize, compute, and extract information, just a tool to help the staff to reduce the business-like work, so that the time of process planning is saved. These systems are not universal because they do not have the decision-making ability and the sort function.

The first true CAPP system, which had the universal significance, was launched by a Norway company in 1969, named as AUTO PROS. From then on a lot of the CAPP systems are subject to the impact of this system. The commercial AUTO PROS system was officially issued in 1973.

Another important system was the CAM-I's automated process planning system that was introduced by the computer-aided manufacturing International (CAM-I) in the United States in 1976. It was considered as a landmark or milestone in the history of CAPP development. It is a relatively long process of the development of CAPP system. From then on, the research of CAPP has gotten a very huge development, from the search logic structure, variant, generative, and hybrid to expert system. Now there is a large number of CAPP prototype systems and commercialization CAPP systems.

As the design process is supported by many computer-aided tools, CAPP has evolved to simplify and improve process planning and achieve a more effective use of manufacturing resources.

In early days, many CAPP systems developed were based on the variant approach. Then since the 1980s, the generative approach and the semigenerative approaches were adopted. Artificial intelligence techniques were introduced in CAPP usually entitled as either knowledge-based or expert systems. A large number of industrial companies had acquired CAPP systems for integration of design and manufacturing and to compensate for the shortage of skilled process planners.

Manufacturers have been pursuing an evolutionary path to improve and computerize process planning in the following six stages:

Stage I – Manual classification; standardized process plans
Stage II – Computer maintained process plans
Stage III – Variant CAPP
Stage IV – Generative CAPP
Stage V – Dynamic, generative CAPP
Stage VI – Hybrid CAPP and expert systems

Prior to CAPP, manufacturers attempted to overcome the problems of manual process planning by basic classification of parts into families and developing somewhat standardized process plans for parts families (Stage I). When a new part was introduced, the process plan for that family would be manually retrieved, marked-up, and retyped. Although the productivity was improved, the quality of the planning of processes was not improved and the differences between parts in a family were not taken into account. There was no improvements in production processes.

Initially, CAPP evolved as a means to electronically store a process plan once it was created, retrieve it, modify it for a new part, and print the plan (Stage II). All the process plans were maintained with computers. Other capabilities of this stage are table-driven cost and standard estimating systems.

This initial CAPP evolved into what is now known as variant CAPP. However, variant CAPP is based on a group technology (GT) coding and classification approach to identify a larger number of part attributes or parameters. These attributes allow the system to select a baseline process plan for the part family and accomplish about 90% of the planning work. The planner will add the remaining 10% of the effort modifying or fine-tuning the process plan. The baseline process plans are manually entered and stored in the computer. They were considered as a super planner, that is, developing standardized plans based on the accumulated experience and knowledge of multiple planners and manufacturing engineers (Stage III).

The next stage of evolution is generative CAPP (Stage IV). At this stage, process planning decision rules are built into the system. These decision rules will operate based on a part's GT or feature-based technology coding to produce a process plan that will require minimal manual interaction and modification.

While CAPP systems are moving more and more toward being generative, a pure generative system that can produce a complete process plan from part classification and other design data is a goal of the future. This type of pure generative system (in Stage V) will involve the use of artificial intelligence type capabilities to produce process plans as well as be fully integrated in a CIM environment. A further step in this stage is dynamic generative CAPP that would consider plant and machine capacities, tooling availability, work center and equipment loads, equipment status, and maintenance downtime in developing process plans.

The process plans developed with a CAPP system at Stage V would vary over time depending on the resources and workload in the factory. For example, if a primary work center for an operation is overloaded, the generative planning process would evaluate work to be released involving that work center, alternate processes, and the related routings. The decision rules would result in process plans that would reduce the overloading on the primary work center by using an alternate routing that would have the least cost impact.

Dynamic generative CAPP also implies the need for online display of the process plan on a work order–oriented basis to insure that the appropriate process plan is provided to the shop floor. Tight integration with a MRP-II system is needed to track shop floor status and load data and assess alternate routings according to the schedule. Finally, this stage of CAPP would directly feed shop floor equipment controllers or, in a less-automated environment, display assembly drawings online in conjunction with process plans.

The intelligent hybrid CAPP systems at Stage VI use artificial intelligence, neutral network, and other modern advanced algorithms to realize the generation of optimal process plan and easily integrated with computer-aided design (CAD)/CAM. They often combine more than two approaches to enforce the capability of the CAPP systems. And even more, the CAPP system might be developed as an expert system, or even a smart system.

3.4 Classification of CAPP systems

Today, the CAPP systems can be classified into five types: variant CAPP systems, generative CAPP systems, expert system-based CAPP system, neutral network-based CAPP system, and hybrid CAPP systems.

3.4.1 Variant CAPP

The system logic involved in establishing a variant CAPP system is relatively straight forward. It matches a code with a similar code that is related to a pre-established process plan maintained in the system. The initial challenge is how to develop the GT classification and coding structure for the part families and how to manually develop a standard baseline process plan for each part family.

The variant CAPP approach is based on the computerized extension of the manual approach to process planning. It is based on the concept that parts possessing similar manufacturing characteristics have similar process plans. In this approach, parts in the company are grouped into part families in accordance with their manufacturing characteristics by using a part classification and coding scheme. Hence, parts with similar manufacturing characteristics form a part family. The various parts to be manufactured are classified, given codes, and grouped within certain families. A standard process plan is then established for each part family and stored in a database. The whole process plan documents the operations as well as the sequence of operations on different machines. The process plan for a new part is thus created by identifying and retrieving an existing standard process plan for a group of similar parts, and if necessary, editing this standard process plan to suit the new part. A code number is used for each family of parts to permit efficient retrieval of the appropriate standard process plan for a new part. The similarities in design attributes and manufacturing methods are exploited for the purpose of forming a part family. The variant CAPP system offers lots of flexibility as one can do lots of editing and changes as per the requirements.

Group technology

GT is a methodology that seeks to take advantage of the design and manufacturing similarities among the parts to be produced by coding and classifying the parts into families.

One company found that by disassembling each product into its individual components and then identifying the similar parts, 90% of the 2,000 parts fell into only five major families. For example, a pump can be broken down into its basic components, such as the motor, housing, shaft, flanges, and seals. Note that, in spite of the variety of pumps the company manufactures, each of these components is basically the same in terms of its design and manufacturing characteristics. Consequently, all housings can be placed in one family of housings, and so on. GT becomes especially attractive because of the ever-greater variety of products, which are often produced in batches. Since nearly 75% of manufacturing today is batch product, it is very important to improve the efficiency of batch production.

Part family coding systems

Three major industrial coding systems are described here:

1. MICLASS System (the Netherlands)

MICLASS stands for Metal Institute Classification System and was developed by TNO, the Netherlands Organization for Applied Scientific Research in the early 1970s. Now it was renamed as MUTICLASS. The MICLASS system was developed to help automate and standardize several design, production, and management functions. It involves up to 30 digits. This system is used interactively with a computer that asks the user a number of questions. On the basis of the answers given, the computer automatically assigns a code number to the part. The typical MICLASS code for a machined part is listed in Table 3.1.

2. The KK-3 system

The KK-3 system is a general-purpose system for parts that are to be machined. It uses a 21-digit decimal system. This code classifies dimensions and dimensional ratios, such as the length-to-diameter ratio of the part. The structure of a KK-3 system for rotatory components is listed in Table 3.2.

3. Opitz coding system (Germany)

The Opitz coding system was the first comprehensive coding system developed, named after a Germany professor, Herry Opitz (1905–1977). An Opitz code consists

Table 3.1: Typical MICLASS code for a machined part.

Pos.	Code	Items	Meaning
0	1	Prefix code	Machined parts/sheet metal code
1	2		Round part, all diameters visible from one end
2	7	Basic form	One machined OD visible from each end and a groove
3	0		No inside diameter
4	0	General manufacturing operations	No secondary elements (holes, slots, flats, curved faces, etc.)
5	4	Function	Axle
6	1		
7	1		Outside diameter range:0.251–0.371 in.
8	0		
9	5	Dimensions	Length:12.51–13.00 in.
10	7		
11	0		Inside diameter: non
12	0		
13	8	Tolerances	Geometric concentricity tolerance <0.01
14	2	Material chemistry	High strength steel
15	8		
16	2	Material shape	Round bar stock
17	1	Production quantity	26–57
18	0	Tool axis complexity	No secondary element orientation
19	3		Thread diameter 5/16 in
20	5		Thread per inch: 18
...	...	User defined	...
30	...		...

Source: Courtesy of Organization for Industrial Research.

Table 3.2: The structure of a KK-3 system for rotational components.

Digit	Items (Rotational component)		
1	Part name		General classification
2			Detail classification
3	Materials		General classification
4			Detail classification
5	Major dimensions		Length
6			Diameter
7	Primary shapes and ratio of major dimensions		
8	Shape details and kinds of processes	External surface	External surface and outer primary shape
9			Concentric screw-threaded parts
10			Functional cut-off parts
11			Extraordinary shaped parts
12			Forming
13			Cylindrical surface
14		Internal surface	Internal primary shape
15			Internal curved surface
16			Internal flat and cylindrical surface
17		End surface	
18		Nonconcentric holes	Regularly located holes
19			Special holes
20		Noncutting process	
21	Accuracy		

Source: Courtesy of Japan Society for the Promotion of Machine Industry.

basically of nine digits (123456789) representing design and manufacturing data, with 1–5 describing the shape of the part and 6–9 describing part dimension, precision, and material. The number can be extended with four additional codes if necessary to express the type and sequence of production operations to be performed. The structure of Optiz code system is shown in Figure 3.2.

The code for a rotating part is shown in Figure 3.3. The first digit 0 means that the part has a rotary shape, and the ratio of length to diameter is less than 0.5. The second digit 1 means that the part has a step feature. The third digit 3 means that the part has an inner shape with a step feature. The fourth digit 1 means that the external plane of the part needs to be machined. With these digits, the part is easily classified.

The code for a nonrotating part is shown in Figure 3.4. The first digit 6 means that part is not a rotary part. The second digit 5 means this part is a rectangular part, with

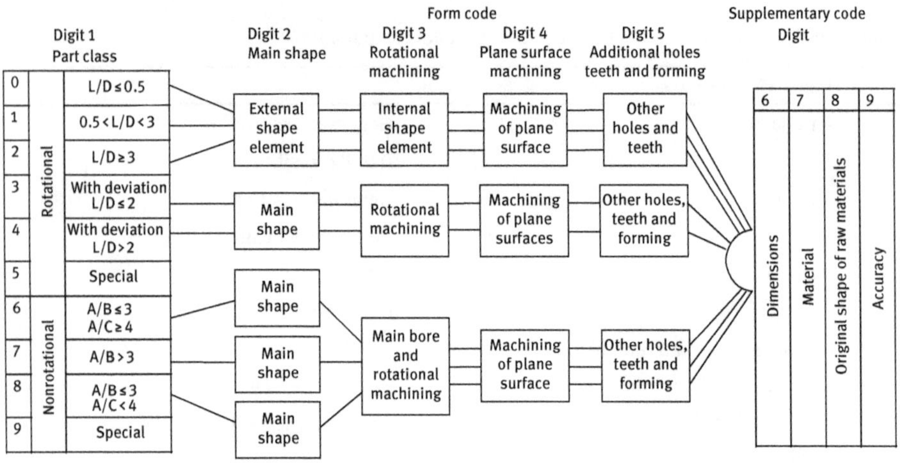

Figure 3.2: The structure of Opitz coding system.

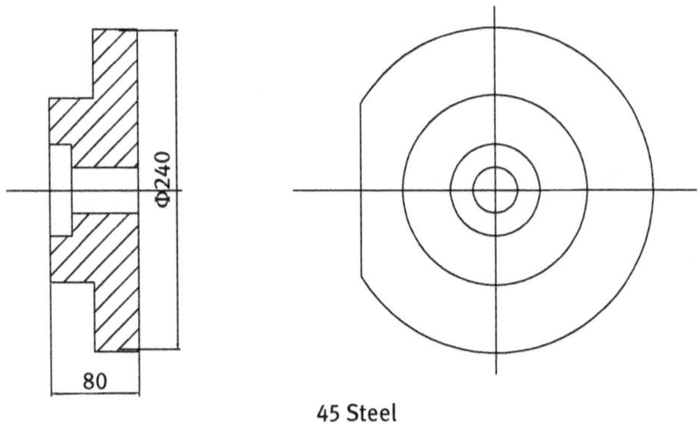

Figure 3.3: Rotating part code: 013124279.

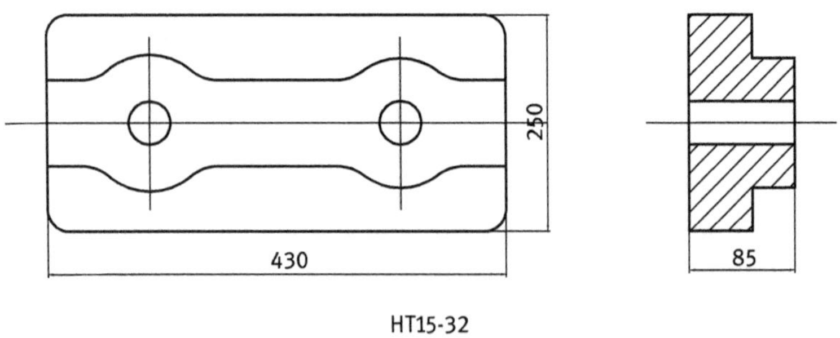

Figure 3.4: Nonrotating part code: 654435078.

a small difference. The third digit 4 means that part has some holes to be machined. The fourth digit 4 means that the part has two planes to be machined.

The variant CAPP systems use GT to classify the targets by comparability (e.g., the part geometry or the process similarity; their structure and function of the components, and so on). Then, the CAPP system designs the process planning for each group and finally establishes the typical process. A standard process plan is designed for the part family. Based on the part family, the process plan can be retrieved and edited for the proper characteristics. The general design process for a GT-based CAPP system can be described as follows:

- Select a part family coding system
- Part family clustering
- Sample part design
- Standard process plan design
- Expression of standard process plan and screening process plan
- Establish a database for standard process plan
- Module design and software programming
- Debugging and using

Feature technology

The variant CAPP system also uses the feature technology, production flow ,analysis and fuzzy clustering methods for the classification of part family. Here, a feature-based CAPP system for manufacturing prismatic parts is presented as an example. The system makes use of form features for component geometry description and process planning functions. Component geometry is described interactively by the user in terms of its shape and features to be machined to create a component database. The system includes the facilities for machine tool selection, set-up planning, tool selection, and operation sequencing. For operation sequencing, feature interaction is taken into account to generate more consistent and optimum process plans. Since system design is based on the integrated database approach, many of the concepts may be applicable to the CAD/CAM systems. The use of form features appears to be providing the missing link between CAD and CAM systems. For many years, the CAM-I system has been trying to categorize part features and the report includes a comprehensive list of features for three classes of components, namely, rotational, prismatic, and sheet metal parts and describes the data structure for all the features.

The introduction of feature-based systems has sparked the enthusiasm of engineers and manufacturers because they potentially allow both design and manufacturing engineers to perform their separate tasks using a language they have always shared, the language of features – holes, slots, pockets, and so on. The use of a common language of features promotes the tighter integration of design and manufacturing functions and makes it easier for design and manufacturing engineers to master and manipulate them than the arcs, lines, and splines that characterize the traditional CAD/CAM systems.

3.4.2 Generative CAPP

The generative CAPP system is designed by the logic of the decision-making process and algorithms. It is an automatically generated process. With the new technology of artificial intelligence, the knowledge-based CAPP expert systems, the CAPP tooling system and framework, and the neural network-based CAPP systems are developed.

The first key to implement a generative CAPP system is the development of decision rules appropriate for the items to be processed. These decision rules are specified using decision trees, computer languages involving logical "if–then" type statements, or artificial intelligence approaches with object-oriented programming.

The nature of the parts will affect the complexity of the decision rules for generative planning and ultimately the degree of success in implementing the generative CAPP system. The majority of generative CAPP systems implemented to date have focused on process planning for fabrication of sheet metal parts and less complex machined parts. In addition, there has been significant recent effort with generative process planning for assembly operations, including printed circuit board (PCB) assembly.

A second key to a generative process planning is the available data related to the part to drive the planning. This approach would involve a user responding to a series of questions about the part information. It can also interpret the design data from CAD directly to get the manufacturing data.

In the generative process planning systems, new plan is made automatically from scratch for each part using the computers, without involving human assistance. The computer program uses a set of algorithms that enables it to take a number of technical and logical decisions to attain the optimum final manufacturing process plan.

One has to give certain inputs to the systems like detailed description of the part to be manufactured. This CAPP system synthesizes the design of the optimum process sequence based on the analysis of the part geometry, material, and other relevant parameters.

The design process for a generative CAPP system can be described as follows:
- Input part information or CAD data directly
- Choose proper machine methods
- Generate main process plan
- Modify and expand the main process plan
- Operation design
- Output process plan

Determination of machining methods can refer to the precision and surface roughness, as listed in Table 3.3. Here, the logical if–then statements can be listed as follows:

Table 3.3: Machining methods for external cylindrical surface feature.

No.	Machining chains	Economic precision	Roughness µm (Ra)	Scope of application
1	Rough turning	IT11– IT12	> 10–80	Metals except for
2	Rough turning–semi-finish turning	IT9	> 2.5–10	quenched steel
3	Rough turning–semifinish turning–finishing turning	IT7– IT8	> 0.63–2.5	
4	Rough turning–semifinish turning–finishing turning–polishing/rolling	IT7–IT8	> 0.02–0.63	
5	Rough turning–semifinish turning–grinding	IT7–IT8	> 0.03–1.25	For both
6	Rough turning–semifinish turning–rough grinding–finishing grinding	IT6–IT7	> 0.08–0.63	quenched steel and unquenched
7	Rough turning–semifinish turning–rough grinding–finishing grinding–honing	IT6	> 0.01–0.16	steel, not for nonferrous metal.
8	Rough turning–semifinish turning–finishing turning–diamond turning	IT5–IT6	>0.02–0.63	Nonferrous metal
9	Rough turning–semifinish turning–rough grinding–finishing grinding–mirror quality grinding	–IT6	<0.01–0.04	With super precision
10	Rough turning–semifinish turning–rough grinding–finishing grinding–honing	IT6	<0.01–0.16	

If the material is nonferrous metal, then process 8
If the surface roughness is less than 0.63, then process 4
If the surface roughness is less than 0.63, then process 6
If the surface roughness is less than 0.63, then process 8
If the economic precision is IT 6, then process 6
If the economic precision is IT 6, then process 7
If the economic precision is IT 6, then process 8
If the economic precision is IT 6, then process 9
If process 8, then rough turning–semifinish turning–finishing turning–diamond turning.

3.4.3 Expert system-based approach for CAPP

An expert system is an intelligent computer program that can understand problem-specific knowledge and uses knowledge and interference procedure to suggest alternative solutions for solving problems of a similar nature. Such a system is particularly effective for problems with a limited knowledge domain. Since the preparation of process plans in the manufacturing environment is a very complex task that requires a considerable amount of human expertise, the expert system-based approach thus has great potential for research related to process planning development. The expert

CAPP system is based on the artificial intelligence. It is an automatically generated process, just like the generative CAPP system. But the generative CAPP system depends on the logic of the decision-making process and algorithm, while the expert CAPP system is on the repository and inference engine.

An expert system mainly consists of the following components:

User interface. This provides communication channels between the users and the internal expert system.

Interference engine. This module controls the interference procedure. It uses a certain control strategy with the help of procedural knowledge and built-in search algorithms.

Knowledge base. The knowledge base stores the domain knowledge, which is used by the interference engine for achieving the user predefined goals.

Knowledge acquisition module. The function of this module is to provide the user or knowledge engineer of the system a mechanism for updating the knowledge base as a result of the dynamic variation of the domain environment.

Explanation facility. This module allows the system to explain its behavior. The purpose of the explanation facility is therefore twofold: debugging the knowledge in a development environment and training the nonexpert.

The working procedure of an expert process planning system normally can be stated as follows:

(1) Receive part information that includes nominal shape, form feature, surface, dimension and tolerance, and material.
(2) Using the operation selection rules in the knowledge base, the process inference engine begins to select the manufacturing operations according to the manufacturing features in the part information database.
(3) The process inference engine separates the manufacturing operations into various process routes and adds the related auxiliary processes such as heat treatment processes and workbench processes to the process route. This is done on the basis of the process sequencing rules, the accuracy of the selected operations, the nominal shape, and the material information. Machine tools are also selected for these process routes.
(4) After the operations have been sequenced and the process routes have been formed, the process inference engine deals with the setups within each of the process routes according to the global dimension and tolerance information among the manufacturing features by using the set-up rules. This includes set-up methods, selection of fixture datum and fixture/jigs tools, and the sequencing of operations within each setup.
(5) The process inference engine selects the cutting tools and gauge tools and also determines the manufacturing parameters for each operation.

The final process plan is then output to the user through the user interface.

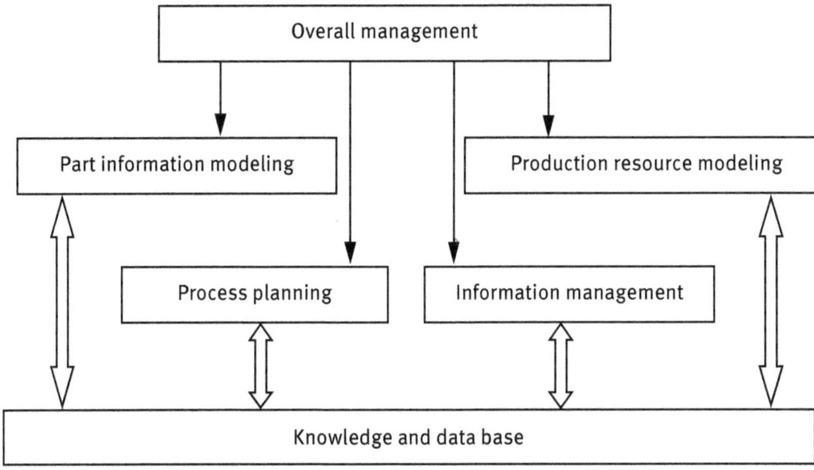

Figure 3.5: The architecture of expert system-based CAPP system.

The architecture of the expert system-based CAPP system consists of four components, namely, the part information modeling module, the production resource modeling module, the process planning module, and the information management and maintenance module.

All these modules are controlled by the overall management. The information flow of the CAPP system is shown in Figure 3.5.

Part information modeling establishes and outputs the process-oriented part information in the form of a neutral file and includes the nominal data of shape, form feature, dimension and tolerance, surface information, and material of the input part.

Production resource modeling provides a mechanism for the users to initiate the CAPP system with the production resource information and process knowledge, and includes machine tools and their manufacturing capabilities, manufacturing operations, cutting tools and their cutting capabilities, part material initialization, blank information input, manufacturing operation selection rules, cutting parameter calculation rules, machine allowance determination rules, and so on.

The process planning performs the function of generating the numerical control (NC)-oriented process plan for the input part by using the process knowledge as well as the input data of the part and the production resource data.

The expert system-based approach for CAPP can automatically acquire knowledge to a certain extent by using the inference rules in the process planning domain. It can also explain its inference procedure to the users. It has the intelligence of knowledge base and is partially adaptive to the dynamic manufacturing environment.

3.4.4 Neural networks-based approach for CAPP

Neural networks-based approach is an effective means for intelligent manufacturing systems. A neural network is an abstract simulation of a real nervous system that contains a collection of neurons communicating with one another. A neural network is defined as a massively parallel interconnected network of simple (usually adaptive) elements and their hierarchical organizations, which are intended to interact with the objects of the real world in the same way as biological nervous systems do. A neural network attempts to achieve the best goal through a dense mesh of computing nodes and connections.

In comparison with expert systems, neural networks have the following advantages: relatively robust with respect to fabrication defects and difficult failures, high processing speed through massive parallelism, and the ability to adapt to the dynamic manufacturing environment because of efficient knowledge acquisition capability. However, they have not been widely adopted as an alternative stand-alone approach for developing CAPP systems. This is due to the following deficiencies:

- A neural network cannot explain its results explicitly, which implies that the user interface of a neural network may not be as user friendly or productive as that of an expert system.
- The knowledge representation of a neural network is vague and cannot be easily understood.
- The configuration of a neural network is usually time consuming. A trial-and-error method has to be used to find the proper neural network architecture for a given problem.
- The current neural network learning algorithms are not efficient enough and cannot guarantee the convergence.
- Lack of systematic and efficient methods for identifying an appropriate training set for a specific application.
- For real-life applications, it is inevitable that the neural network will have to learn a significant volume of declarative knowledge. The neural network, in this case, may need a large number of nodes or a complex structure. Under these circumstances, this requirement aggravates the difficulty in training the neural network. The problem may be more easily dealt with by using the conventional knowledge engineering techniques.

Hence, in spite of the fact that the neural networks possess many good features, the above deficiencies clearly indicate that the technology is complementary rather than competitive with traditional artificial intelligence approaches such as expert systems for developing CAPP systems.

3.4.5 Hybrid CAPP system

Hybrid CAPP system had the meaning of the combination of the variant system and generative system at early days. Some parts of the process plan are created by identifying part families and retrieving an existing standard process plan, while other parts of the process plan are generated based on the analysis of the material, part geometry, and other technical aspects that would affect the manufacturing process.

With the development of computer science and artificial intelligence, some knowledge-based CAPP system, neural-based approach for CAPP, and expert system-based approach for CAPP systems have been developed. Thus, hybrid CAPP system becomes the combination of any newly developed approaches. Here, the hybrid expert- and neural-based approach for CAPP is taken as an example to explain how the two approaches are combined, and its functional models are described.

The hybrid expert- and neural-based approach for CAPP

In order to develop an effective process plan, the process planners are required to process a lot of information and knowledge. Such information and knowledge are usually represented in a form closely resembling a rule-based expert system. Hence, if a computer system is to simulate the intelligence of a process planner in developing an effective process plan, it must be equipped with the capability of processing both heuristic knowledge and imprecise decisions domain knowledge. The computer systems must also be able to learn and thus improve its performance when experience in developing process plans has been accumulated. Hence, the hybrid expert and neural approach represents an effective method for knowledge integration and management in the design of intelligent CAPP systems in the CIM environment. Indeed, expert systems and neural networks are complementary. Hybrid expert and neural architectures are developed for CAPP applications. Nevertheless, the advantages of combining expert systems and neural networks to form a hybrid system for process planning are as follows:

(1) Reduce the effort in knowledge acquisition greatly since there are fewer rules in the system. Consequently, it is easier to maintain the system throughout its life cycle.
(2) Update new knowledge automatically by presenting new examples to the neural network.
(3) The extraction of implicit knowledge is relatively easy by the help of a certain amount of explicit knowledge.
(4) Neural networks can preprocess the application data into a form that can be effectively handled by the expert system.
(5) For complex problems represented by multinetworks, the expert systems can serve as the system administrator to control the information and knowledge flow

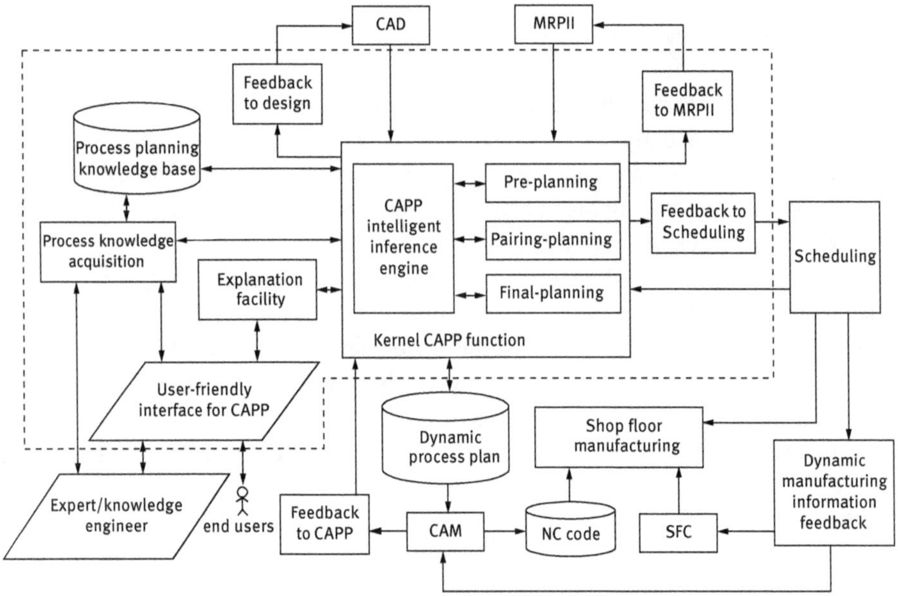

Figure 3.6: Hybrid CAPP intelligent functions architecture.

through these networks to improve the performance of the neural networks. The expert system also provides an explanation facility for the system.

(6) The expert system and neural networks can work together in such a way that the expert system analyze the responses from the neural network and processes them to permit easy interpretation of the results. This analysis may also lead to new training patterns for the neural networks. This improves the overall performance of the system.

A hybrid expert- and neural-based function model for CAPP

The architecture of an integrated CAPP intelligent function model is presented in Figure 3.6. It can be noted that the CAPP function is dynamically and bidirectionally integrated with other functions such as CAD, CAM, MRP-II, scheduling, and shop floor control. The central features of the architecture are the CAPP intelligent function model.

The CAD function model

The CAD function model consists of the kernel of Parasolid or ACIS, and the STEP toolkit, all CAD functions, and one database for storing the process planning-oriented part information. The functions of CAD just like any other CAD systems, such as geometry modelling etc. But here the CAD function can also create the process planning-oriented product design data by using the manufacturing features and related specifications. The STEP toolkit provides a mechanism to apply the PDES/STEP-integrated resource/application protocol to further develop the CAD system. In so doing, the final CAD system can communicate easily with other systems, which also use PDES/STEP in the manufacturing environment.

Product data are represented in EXPRESS language, which can be mapped into the head file of any executable program such as C++ head file with the EXPRESS-mapping processor. This function helps the programmer to develop the CAD system with a unified data structure. In addition, product data can also be mapped into a suitable storing format of certain database software commercially available for every product data represented. This mapping processor guarantees that the data structure in the head file is identical to the structure of the data stored in the database. This will facilitate the input of product data to the database through the standard data access interface (SDAI), which is a PDES/STEP implementation method. In addition, the CAPP system can communicate with the CAD database conveniently by using the SDAI function. Similarly, by using the SDAI function, both the MRP-II system and the CAPP system can communicate easily with the manufacturing resource information. Indeed, the SDAI function helps a lot to integrate the various systems.

The CAPP intelligent function model

The CAPP intelligent function model mainly consists of databases, auxiliary intelligent functions, information feedback, and kernel CAPP function.

Databases: There are two types of databases in the model, the process planning knowledge database and the process plan results database. The process planning knowledge database is a hybrid knowledge database, which includes not only the facts and logical rules but also the heuristic rules in terms of neural network. The process plan results database stores the final process plans.

Auxiliary intelligent functions: The auxiliary intelligent functions comprise the process knowledge acquisition, the explanation facility, and the user-friendly interface for the CAPP function.

Feedback from CAPP: The feedback from CAPP is the main feedback to design function, feedback to MRP-II function, and feedback to the scheduling function. The feedback to design includes the technical feasibility of the manufacturing features, the technical feasibility of the entire component and its cost of manufacturing, and the need to adopt alternative manufacturing features. The feedback to MRP-II comprises the evaluation of the vendor's qualifications and the changes in the shop floor that include changes of machine tools, fixtures/jigs, cutting tools, and gauge tools. Feedback to scheduling is the information to facilitate scheduling or rescheduling of the manufacturing operations in view of the current process plan produced by the CAPP system.

Kernel CAPP function: The kernel CAPP function consists of CAPP intelligent inference engine, preplanning, pairing planning, and final planning. The CAPP intelligent inference engine plays the key role in serving as a channel of communication between the kernel CAPP subfunctions and the CAPP knowledge base. It also controls the implementation strategy of the kernel subfunctions, which are preplanning, pairing planning, and final planning. These subfunctions are hierarchically separated by using the hierarchical planning method as it is carried out by the human expert in developing process plans. They also have child functions as described as follows:

(1) **Preplanning:** This subfunction includes manufacturing feature analysis, feature relationship analysis, dimension and tolerance analysis, and manufacturing operation selection. These child functions are performed in an order. The purpose of the preplanning function is therefore to analyze the entire component information, especially the manufacturing features, and select the manufacturing operations according to the manufacturing resources currently available.

(2) **Pairing planning:** This subfunction includes process route sequencing, machine tool selection, determination of stock removals, set-up planning, process dimensions determination, and process drawings generation. Therefore, the role of this subfunction is to sequence the process route, select the machine tools, determine the allowance for each process, calculate the process dimensions, generate the process drawings, and perform set-up planning. All these activities are carried out on the basis of the design data provided by the CAD system, information concerning availability of machine tools and fixtures/jigs from the MPR-II system, and schedule constraints from the scheduling system.

(3) **Final-planning:** This subfunction consists of cutting tool selection, cutting parameter determination, gauge tool selection, alternative process plans selection, and NC-oriented process plan data editor. All these functions are performed in accordance with the constraints on availability of cutting tools as well as gauge tools, and the requirements specified by the process planner.

The CAPP intelligent inference engine control strategy

The CAPP intelligent inference engine control strategy is used to manage, control, and coordinate the inference strategy used by the CAPP subfunctions. Two main kinds of inference strategies are used in CAPP, that is, the expert inference strategy and the neural network inference strategy. These two inference strategies can be used separately during the process of developing process plans. The mechanism of the expert inference strategy corresponds to the expert inference engine. The mechanism of neural network inference strategy corresponds to the parallel calculation by using prestored trained weights. The expert inference strategy is used for handling the rule-based knowledge, while the neural network inference strategy is used for handling the heuristic knowledge and imprecise decisions domain knowledge. Hence, when knowledge cannot be represented easily in the form of rules, it will be represented by using a neural network and accordingly the neural network inference strategy will be used, or the expert inference strategy will be used. The CAPP inference engine control strategy, therefore, follows the procedure presented below.

The CAPP system reads design data of the part from CAD systems. The design data are compared with the design data of existing parts, which have been used to produce process plans by the CAPP system and are now stored in the process plan database. The CAPP system judges whether the design data are familiar, strange, or similar to the old design data. If the new design data are strange to the CAPP system, the inference engine uses the expert system inference strategy. After the final process plan is obtained, the neural network is trained by using both the new design data as input and the final process plan data as output to improve the CAPP systems' cognitive degree for this new part. If the new design data are strange to the CAPP system, this implies that little or no training has been conducted to the neural network by using the new design data. Hence, the neural network has acquired very little knowledge about the new design data. The expert system inference strategy should therefore be used in developing the process plan for this new part. If the design data are familiar to the CAPP system, the CAPP system uses the neural network inference strategy to realize the inference procedure. Because of the neural network's parallel calculation, it takes a relatively short time to obtain the NC-oriented process plan. Actually, if familiar to the CAPP system, it implies that the neural network has already acquired enough knowledge about the new design data. If the design data are judged to be similar to the old design data by the CAPP system, the CAPP system allows the user to select the inference strategy to be used.

Comparison of the advantages of the hybrid expert- and neural-based approach and the other approaches to CAPP is summarized in Figure 3.7. Obviously, the hybrid expert- and neural-based approach possesses better characteristics than other

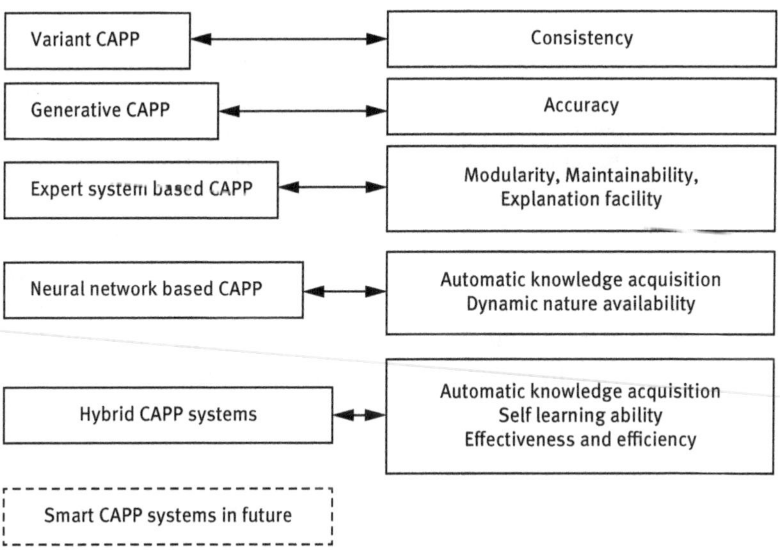

Figure 3.7: Comparison of the CAPP approaches.

approaches. In addition, since the system has self-learning ability, its performance improves as more knowledge has been acquired. Therefore, the hybrid expert and neural function model forms a very appropriate basis for developing CAPP systems under the dynamic manufacturing environment.

3.5 Methods/technologies of CAPP

There are ten established methods/technologies of CAPP in the development of CAPP systems. They are, namely, feature-based technology, knowledge-based systems, neural networks, genetic algorithms (GAs), agent-based technologies, fuzzy set theory and fuzzy logic, Petri nets (PNs), Internet-based technologies, STEP-compliant method, and functional blocks (FB). They are used to overcome the issues of machining operation, tool and machine selection and sequencing, feature extraction, reorganization, interpretation and representation, knowledge integration, representation, acquisition and sharing, set-up planning, energy consumption, linear and nonlinear planning, integration of product and manufacturing data, intelligent tool path generation, optimization problems, intelligent decision-making and sharing of knowledge, integration of process planning and scheduling, and so on.

Feature technology plays a key role in process planning with feature recognition and design by features. It has been adopted by many process planning systems, due to its ability to facilitate the representation of various types of part data in a significant form to drive automated CAPP. Knowledge-based technology allows the capturing of knowledge from experts and is able to simulate the problem-solving skills of a human expert in a particular field. Neural networks have an ability to recognize transitional and complex features without feeding any pervious knowledge into the system, and they also have the ability to derive rules or knowledge through training with examples and can allow exceptions and irregularities in the knowledge. GAs are generally good at finding feasible solutions to problems quickly. They do not need gradient information or smooth functions, and they have to be carefully structured and coded. Agent-based approach offers some unique functionality for distributed product design and manufacturing. The fact that process planning for a complex part can be broken down into smaller planning problems makes these problems manageable by a number of intelligent agents working in cycle. Fuzzy logic and PNs technologies are used in CAPP by combining with other technologies to offer better solutions. Internet-based technology enables engineers to achieve the dynamic tool and machine selection, thus aiding in the existing CAPP system to generate realistic and economical process plans. It also allows process planners in any industry to react to any unexpected changes and to support the data exchange between different systems used at different companies. STEP-compliant in CAPP leads to the likelihood of using standard data throughout the entire product process chain in the manufacturing atmosphere, hence increasing the system's capability of integrating with other systems in the complete CIM environment. The FBs approach plays an active role in CAPP in response to shop floor uncertainty because FB-embedded algorithms can make decision adaptively to changes at run time and can be integrated with dynamic scheduling and STEP-NC systems.

3.6 Determine dimensions and tolerances

Dimensions are used to describe the size and location of features on parts for manufacture. Dimensions should not be excessive, either through duplication or dimensioning a feature more than one way. Size dimension might be the overall width of the part or the diameter of a drilled hole. Location dimension might be the length from the edge of the object or to the center of the drilled hole.

Tolerance is the total amount a dimension that may vary and is the difference between the maximum and minimum limits. Tolerances are used to control the amount of variation inherent in all manufactured parts. In particular, tolerances are assigned to mating parts in an assembly.

The dimensions and tolerances must be determined for each machining operation in the process planning. Usually, a dimensional chain is formed to calculate the tolerance stack-up, and to determine the dimensions of each work operations. Here, the tolerance analysis and tolerance synthesis approaches are used to determine the dimensions and tolerances.

Tolerance analysis

Tolerance analysis is the general term for activities related to the study of potential accumulated variation in mechanical parts and assemblies. Engineers analyze tolerances for the purpose of evaluating geometric dimensioning and tolerancing. Methods include 2D tolerance stacks, 3D Monte Carlo simulations, and datum conversions.

PTC Creo Tolerance Analysis Extension (TAE), powered by CETOL Technology, is an extension to PTC Creo Parametric. With PTC Creo TAE, the impact of tolerances and dimensioning on product designs can be evaluated before the product reaches either prototyping or manufacturing. PTC Creo TAE provides two analysis tools: worst-case analysis and statistical analysis. Worst-case analysis, often used for critical components, considers what happens when each of the components in the stack-up reaches the maximum acceptable measurement. In statistical analysis, each component's variation is modeled as a statistical distribution. These distributions then are added together to predict the distribution of the assembly measurement.

Using these tools, designers can produce products that meet customer requirements within acceptable margins. The results are shorter product development cycles, lower product cost, and higher product quality.

Tolerance synthesis

Tolerance synthesis is a procedure that distributes assembly tolerances between components or distributes final part design tolerances between related tolerances. Making a product conform to the functional requirements indicated by the customer supposes to be able to manage the manufacturing process chosen to realize the parts. A simulation step is generally performed to verify that the expected generated deviations fit with these requirements. It is then necessary to assess the actual deviations of the process in progress. This is usually done by the verification of the conformity of the workpiece to manufacturing tolerances at the end of each setup. It is thus necessary to determine these manufacturing tolerances. This step is called manufacturing tolerance synthesis. The manufacturing tolerance synthesis is used to determine the working tolerances for the working dimensions. It is a necessary step in the process planning.

3.7 Design fixtures and jigs

Jigs and fixtures are special purpose tools that are used in machining, assembling, and inspection operations. It is one of key problems for process planning. They are used to position, clamp, and protect components and subassemblies at all stages of the manufacturing process. New jigs and fixtures must be designed, manufactured, and deployed. Jigs and fixtures are commonly fabricated by machining metal, wood, or plastic. When this work is done in-house, it can tie up machine tools used for production work. When the work is subcontracted, lead times can take weeks. Whether they are made in-house or outsourced, jigs and fixtures are subject to the same design-for manufacturability limitations that govern production work. This can result in designs that are not optimal for the operation. For the process planning, each operation needs a set of fixture to locate and clamping the workpiece. Some fixtures are traditional and commonly used, and some are special for the workpiece. When a large amount of the product to be produced, a special set of fixture must be designed and manufactured. So the design fixture and jigs are very important in CAPP systems.

3.8 Some commercial CAPP systems

According to a survey on CAPP by Yusri Yusof from Malaysia, more than 100 CAPP systems exist. They are written in various programming languages such as C++, LISP, Turbo Pascal, Fortran, Prolog, and dBase. Sample variant systems are CAMSS, GLM, and PerMIA. Sample generative systems are AFR, ALPS, AMOPPS, CROPS, ESTPAR, FBD, DCAPP/SM, GEOPDE, GFAS, GLM, IKOOP, KAPLAN, K-base, MCOES, PART,

ROBEX, RATE, and TVCAPP. Sample hybrid systems are CADEXCAP, COMPLAN, RDCAPP, SMT, and TAMCAM.

There are some commercial CAPP systems available in current markets. The AUTO PROS (Norway, 1966) was the first commercial CAPP package available in the market. CAM I's Automated Process Planning (USA, 1976) was developed in the United States and became popular CAPP systems in the early days. In China, HZ-RCAP (China, 1986) was developed for the rotational parts using GT. NH-CAPP (China, 1988) was developed for both the rotational parts and the nonrotational part, such as shaft and gear housing. These CAPP systems were used by some big companies. Nowadays, there are several Chinese CAPP systems available in the market, such as the TianYu CAPP from WuHan Tianyu Software Co. Ltd., CAXA CAPP from Beijing CAXA technology Co. Ltd, and KM CAPP from Wuhan KM Co. Ltd.

From Yusri Yusof's survey, it has been found that most of the CAPP work carried out on machining manufacturing resolve the problem issues of operation, tool and machine selection and sequencing, feature extraction, reorganization, interpretation and representation, knowledge integration, representation, acquisition and sharing, set-up planning, energy consumption, linear and nonlinear planning, integration of product and manufacturing data, intelligent tool path generation, optimization problems, intelligent decision making and sharing of knowledge, integration of process planning and scheduling, and so on. From the survey of CAPP of the last 12 years, it is found that during the 2002–2007 period of time, there was a pretty balanced work on feature-based, knowledge-based, neural network, GA, Internet-based, and STEP-compliant techniques. However, during the 2008–2013 period, knowledge-based and STEP-compliant technique utilization was increased to some extent.

Here, two CAPP systems are described in detail. They are KM-CAPP and IITD-PP systems.

KM-CAPP

KM-CAPP was the first commercially available CAPP software package in China. It was developed by Wuhan KM Software Company 20 years ago. It has been improving all the time. At present, it dominates the Chinese CAPP market. KM-CAPP has been widely used in automotive, aerospace, and electronic industries.

KM-CAPP has a special objective. Automation is not the only target of CAPP. KM-CAPP has been developed to free the process planner from the fussy and tedious design work. The main objective is to achieve the easy use of operation, the efficient means of process making, the automatic aggregate statistics of the process information, and the information integration with CAD/PDM/ERP systems. Having a good open and integrated port is very important for the research and extension of the CAPP system.

The main function modules of KM-CAPP include process plans management, manufacturing resource management, knowledge-based CAPP system, and interfaces to other systems.

1. Process plans management
   KM-CAPP provides a workbench for editing process plans, mainly to edit the forms for routings, standard process plan, manufacturing resource management, formulae calculation, and operational engineering drawing.
2. Customerize tools
   KM-CAPP permits users to customize their CAPP system. The routing forms can be tailed according the habits of engineers in the company. KM-CAPP utilizes all the technical data and resources including machine tools, tooling, fixture, cutting parameters, and materials. All information is constructed as an object using object-oriented method.
3. Output process plan
   The process plan generated by KM-CAPP should be printed out. The KM-CAPP can preview the output before printing. The output can be printed as an A0–A4 blueprint drawing.
4. Interface with other systems
   KM-CAPP supports many kinds of database, such as Access, SQLServer, and Oracle. The process plan can also be converted into XML format or Word, Excel and Autocad formats. KM-CAPP also has interfaces to CAD, MRP-II, and CAM systems.
5. CAPP to BOM
   BOM from the CAD can be integrated with KM-CAPP. The tree of the BOM can be easily generated in CAPP system.
6. Knowledge-based CAPP system
   KM-CAPP is a knowledge-based CAPP system. The information and knowledge are stored in the system database. The process plan can be generated by using the parameterized methods and knowledge about the manufacturing. The GT, feature-based technology, and knowledge-based are all embedded in the KM-CAPP system.

   KM-CAPP is a powerful CAPP system. What you see is what you get. KM-CAPP also stores the technology information in their cloud database, so it is possible to use KM-CAPP system in cloud manufacturing. The KM-CAPP is a great success to the commercial CAPP systems. It makes the CAPP develop from the laboratory to the market and enterprises. With the KM-CAPP system, thousands of enterprises improve the efficiency of the process planning. And it also promotes the standardization of this technology and the building of enterprise information.

IITD-PP

IITD-PP is another CAPP system introduced by S.K. Gupta. IITD-PP is the name of the system, because it has been developed at Indian Institute of Technology (IIT), Delhi, India. It was a system developed for generating process plans for the prismatic components. Some of the important characteristics of the system are as follows:

1. The system is based on a set of manufacturing features.
2. All modules of the system utilize a central database, which is expressed in terms of manufacturing features.
3. The system is designed in a manner such that it can be easily integrated with other modules in a manufacturing environment for dynamic process planning.
4. The system interacts with the user. Hence, the user gets the necessary assistance for trouble shooting.

IITD-PP software has been made to build a model for part description facilities based on form features. The model supports the required facilities for component geometry description for a typical prismatic component. A user-friendly interface is provided to capture manufacturing-related information at the design stage. Therefore, the model is able to provide an environment for an integrated system, suitable from both design and manufacturing point of view.

The system includes the facilities for set-up planning and provides the opportunity to study the feasibility of automating the set-up planning activities. The scheduling facilities incorporated in the machine tool selection module provide the necessary environment for integrating the module in a CIM environment. The backward planning structure used in the tool selection module makes the tool selection operation easy to understand. By making online communication with the user, this module provides the necessary suggestions and opportunities for making the necessary changes to the design or production facilities, if the component fails to meet the local manufacturing constraints. The operation sequencing module is capable of dealing with feature interaction. This module clearly indicates the importance of feature interaction in operation sequencing, and at the same time indicates the necessity of a feature-based system for handling feature interaction.

At present, the system makes use of a limited number of features to model the typical prismatic components to be machined on a vertical machining center. Set-up planning is done only for the parallelepiped-shaped jobs and the examination of feature interaction is limited to some representative features.

During the past decade, the trend has changed from interfaced automation to integrated automation. The new generation feature-based systems help to bring manufacturing considerations into the design phase, and the coordination and integration between design and manufacturing is partially built-in, and thus the need for the process planning function is partially eliminated. In this respect, the most challenging task is the definition of features which serve both design and manufacturing requirements. As research progresses in this direction, the system will ultimately be completely integrated (design and manufacturing activities will be completely unified), that is, process planning aspects will be part of the design phase itself and the need for separate process planning will be reduced.

At the same time, because of the variability and the complexity in manufacturing environment, it is extremely difficult to achieve a CAPP system when the structure is complex and the components are too much.

3.9 Integration of CAD/CAPP

Today's more advanced CAPP systems take a CAD-based product model as input, which provides the geometric information and dimensions. At best, this is a 3D solid model on which the CAPP system can perform automatic feature recognition. However, some CAPP systems exist that take wire frame models as an input and on which the process planner has to identify the manufacturing features manually. As CAD models often do not contain tolerance, material information, and other additional requirements, some CAPP systems allow for adding this information to the product model manually in order to allow automatic reasoning. Most generative CAPP systems allow for human interaction. A lot of CAPP systems can be classified as semivariant or semigenerative.

Many CAPP systems can handle a feature-based CAD model as input. That is, if they seem to be able to handle such models, the feature-based model first has to be converted into its corresponding solid model representation from which the CAPP system can infer the manufacturing features by means of feature recognition. Manual feature identification as for instance is labor intensive and should be avoided. In the case of feature-based design followed by feature recognition or feature identification, the feature information is first thrown away and recovered later. The design features and the process planning features correspond and can be converted into one another. A feature conversion, however, can be obstructed by different feature representations in CAD and CAPP systems. In fact, in some feature recognition-based CAPP systems, the features are described within the feature recognition algorithms. The definition of new features in many CAPP systems involves the programming of new feature recognition algorithms. Although feature recognition languages can be fairly high-level programming languages, the definition of new features in a CAPP system can involve a lot of work.

3.10 PDM-based CAPP systems

Integrated CAPP systems can be developed based on product data management (PDM). As companies implement PDM and MRP-II systems, the need to address the requirements of manufacturing information management often gets lost in the shuffle. The reality is that the manufacturing information handling function is far more complex and important to the success of a company than top management realizes.

Management tends to assume that PDM and MRP-II can provide all necessary manufacturing information, but in fact neither can provide appropriate access for developing and updating entire process planning packages. For example, MRP-II systems may provide process planning capabilities, but they do not typically have the manufacturing information handling capacity or integration with the graphic capabilities now seen in modern CAPP systems.

PDM databases are developed according to engineering features and therefore are not necessarily relevant to manufacturing considerations. Functionality or part features are the key factors governing the structure of a product-oriented database while the process planner deals with manufacturing elements and processes such as tools and fixtures, turning, milling, hardening, and so on. So when companies suggest that they intend to maintain one overall product database developed from an engineering/design point of view, they overlook manufacturing engineering's need to access to past-manufacturing experience that is so essential to the continuous improvement process.

The key technology, such as the product-oriented process planning, communication between PDM and CAD, the integration between CAPP and MRP-II, version management, workflow management, and configuration management, will help to develop CAPP systems based on PDM.

3.11 Summary

This chapter has made an introduction to CAPP systems. The basic manual process planning approaches of CAPP systems are described. The CAPP systems are classified into variant CAPP, generative CAPP, expert system-based CAPP and neutral network-based CAPP, hybrid CAPP, and so on. The GT-based variant CAPP systems are based on the GT. And the feature-based variant CAPP systems are based on the features of the workpiece. The coding systems for GT are introduced. CAPP has been developed for many years and holds a certain market, but it never enters to the period of the market outbreak, and also have not been widespread promotion and application.

The shortcomings of today's CAPP systems are summarized here. One of the shortcomings of today's commercial CAPP systems is that they do not provide the CAD system, or the designer using the CAD system, with feedback information on cost, manufacturability, and so on. Many researchers have proposed a lot of different ways in which CAD and CAPP systems could become more cooperative. Another shortcoming of present commercial CAPP systems is that they do not communicate with capacity planning functions. Before completing the process plan, the first part of the process plan is derived from information becoming available from feature recognition and set-up selection. Using alternative setups, the jobs are assigned to the resources (loading), based on the actual availability and the actual workload of all the machines in the workshop. Subsequently, the detailed process plan is completed.

Traditional CAPP systems aim to obtain optimal machining processes, machines, and tools capable of performing specified operations from a single job shop having limited available manufacturing resources. However, in a distributed manufacturing environment, there are other available factories capable of performing a task, and it is possible that one of them may provide a more efficient and better process plan. Therefore, CAPP systems can be developed to produce optimal process plans in a distributed manufacturing environment.

With the development of computer science, artificial intelligent, and information technology, smart CAPP systems will be developed in the future.

Questions

1. What are the main tasks of process planning?
2. What is the concept of CAPP?
3. Describe the variant CAPP approach.
4. What is the GT?
5. What is the feature-based technology?
6. Describe the generative CAPP approach.
7. What is the expert system-based CAPP system?
8. What is the hybrid CAPP system?
9. What are the shortcomings of today's commercial CAPP systems?
10. Draw a diagram of the smart CAPP systems in the future.

Tasks

Collect 10 parts from some used products. Use the knowledge in this chapter to classify these parts into different family. Design a standard process plan for each part family.

If you find a new part to be produced, how can you design the process plan for manufacturing the new part?

References

Gupta, S.K., Rao P.N., Tewari N.K., Development of a CAPP system for prismatic parts using feature based design concepts. International Journal of Advanced Manufacturing Technology, 1992, 7: 306–313.

Jain N.K., Jain V.K., Computer aided process planning for agile manufacturing environment, in book Agile Manufacturing: 21st Century Competitive Strategy (Editors A. Gunasekaran), Chapter27, Elsevier Science Publications, New York, 2001: 515–534.

Li L., Fuh J.Y.H., Zhang Y.F., et al. Application of genetic algorithm to computer aided process planning in distributed manufacturing environments. Robotics and Computer Integrated Manufacturing, 2005, 21: 568–578.

Marri H.B., Gunasekaran A., Grieve R.J., Computer aided process planning: A state of art. International Journal of Advanced Manufacturing Technology, 1998, 14: 261–268.
Ming X.G., Mak, K.L, Yan J.Q., et al. A hybrid expert-neural-based function model for CAPP. International Journal of Computer Integrated Manufacturing, 1997, 10(1-4): 105–116.
Yusof Y., Kamran Latif K., Survey on computer aided process planning. International Journal of Advanced Manufacturing Technology, 2014, 75: 77–89.
Zhou, Z.D., Wang, H.Q., Lou P., Intelligent process planning: intelligent CAPP, Chapter 11: Manufacturing Intelligence for Industrial Engineering: Methods for System Self–Organization, Learning, and Adaptation. IGI Global, New York, 2010.

4 Computer aided manufacturing

Questions before you read

1. What is computer aided manufacturing?
2. What is the CNC machining?
3. What are the components of a CNC machine tool?
4. How do you classify the different CNC machine tools?
5. What is the automatic programming tool?
6. Do you know that STEP-NC is a new method for CNC programming?
7. What are the CAM software packages commercially available?
8. Do you know something about 3D printing or additive manufacturingtechnology?

The goal of this chapter is to understand the concept of computer aided manufacturing (CAM). CAM can be understood from both a general point of view and a narrow point of view. Computer numerically controlled machining is described in detail by introducing the numerically controlled machine tools, the tooling systems, and the software systems. The computer numerical control (CNC) program and automatic programming tool (APT) are introduced. Some commercial CAM software packages are also introduced. At the end of the chapter, the 3D-printing and additive manufacturing approaches are discussed.

4.1 Introduction

Computer aided manufacturing is a subsequent computer-aided process after computer aided design (CAD), computer aided engineering (CAE), and computer aided process planning (CAPP). The models generated in CAD and verified in CAE systems can be input into the CAM software, which then controls the machine tool. CAM makes use of computer software and hardware in the conversion of CAD models into manufacturing instructions for NC machine tools.

CAM can be understood from both a general point of view and a narrow point of view. From the general point of view, CAM refers to the use of a computer to assist in all operations of a manufacturing plant, including planning, management, transportation, and storage. The term "CAM" is used to describe a wide range of automation technologies. CAM is the effective use of computer technology in the planning, management, and control of the operations of a manufacturing plant through either direct or indirect computer interface with the plants' production resources. Many computer applications are implemented in a modern manufacturing plant. Computer applications in manufacturing include production scheduling, order release, machine control, material handling control, inventory control, maintaining scheduling, and other tasks. The key aspects of CAM include material requirements planning, capacity planning, facility and material handling device planning, inventory control, tool management, scheduling, quality control, information management, and communication.

https://doi.org/10.1515/9783110573091-004

A broader definition of CAM can include the use of computer applications to define a manufacturing plan for tooling design, CAD model preparation, NC programming, CMM inspection programming, machine tool simulation, or post-processing. The plan is then executed in a production environment, such as direct numerical control (DNC), tool management, CNC machining, or CMM execution.

From the narrow point of view, CAM is known as computer numerically controlled machining. It is just the use of computer software to control machine tools and related machinery in the manufacturing of workpieces. Its primary purpose is to create a faster production process and components and tooling with more precise dimensions and material consistency, which, in some cases, uses only the required amount of raw material (thus minimizing waste), while simultaneously reducing energy consumption.

The numerical control (NC) was the start of the shop floor automation evolution. As U.S. industry emerged from World War II, the NC technology and computers began to shape automation on the shop floor. The development of direct numerical control (DNC) machine tools was a significant leap in productivity. The CNC and NC machines were connected directly to a large mainframe computer that acted as a central program repository and control center. CNC integrated a computer type controller into the machine tool. The CNC machine tool now can store and execute part programs without tapes or a DNC connection to the mainframe computer and with less operator intervention.

As PC-based CAD/CAM software grows more sophisticated, it becomes difficult for a shop to decide exactly what it requires. Because a problem with the program could damage costly machinery and cutting tools, computer simulations may be used to check the program instead of a trial run. CAM commonly refers to the use of NC computer software applications to create detailed instructions (G-code) that drive CNC machine tools for manufacturing parts. Manufacturers in a variety of industries depend on the capabilities of CAM to produce high-quality parts.

Benefits of CAM

The benefits of CAM include a properly defined manufacturing plan that delivers expected results in production.

- CAM systems can relieve the NC programmers from the tedious G-code editing tasks.
- CAM systems can maximize utilization of a full range of production equipment, including high speed, 5-axis, multifunction and turning machines, electrical discharge machining (EDM), and CMM inspection equipment.
- CAM systems can aid in creating, verifying, and optimizing NC programs for optimum machining productivity, as well as automate the creation of shop documentation.
- Advanced CAM systems with product lifecycle management (PLM) integration can provide manufacturing planning and production personnel with data and process management to ensure use of correct data and standard resources.
- CAM and PLM systems can be integrated with DNC systems for delivery and management of files to CNC machines on the shop floor.

4.2 CNC machine tools

4.2.1 Introduction to CNC machine tools

There are various types of machine tools in the world. Some machine tools are used to remove metal or other rigid materials, usually by cutting, boring, grinding, etc. Some machine tools are used to change the form of the sheet material by shearing, bending, or other forms of deformation without removing material. Other machine tools are used to add materials to form a product usually by 3D printing. Whatever type they are, machine tools employ some sort of tool that does the cutting or shaping. All machine tools have some means of constraining the workpiece and provide a guided movement of the parts of the machine. Thus, the relative movement between the workpiece and the tool (which is called the tool path) is controlled or constrained by the machine.

Today, machine tools are typically powered by electrical or hydraulical motors, used to make manufactured components in various ways that include cutting, printing, or certain other kinds of deformation. With their inherent precision, machine tools enabled the economical production of interchangeable parts.

CNC machine tools are automated machining devices that make industrial components without direct human assistance. They use coded instructions that are sent to an internal computer, which allows factories to fabricate parts accurately and quickly. There are many different types of CNC machines, ranging from drills to plasma cutters, so they can be used to make a wide variety of parts.

4.2.2 Principal elements of a CNC machine tool

As there are different types of CNC machine tools, their structures are different from one another. CNC machine tools vary from simple CNC lathes to complex machine centers. However, all machine tools have the principal elements. The principal elements of a CNC machine tool include machine base, spindles, spindle drives, slide drives, and CNC controllers.

Machine base

For many years, casting iron was considered to be the only material suitable for the basic structure of a machine tool. It possessed adequate strength and rigidity and tended to absorb vibration. The use of concrete or ceramics as a machine base is a comparatively new development. The advantages of concrete are its low cost and good damping characteristic. For CNC machine tools, some spot-welding steel plates and reinforcing ribs are used as the machine base. Damping material is inserted between the spot-welding steel plates. The spot-welding structures have good stiffness and can also absorb vibration.

Spindle

In machine tools, a spindle is a rotating axis of the machine, which often has a shaft at its heart. The shaft itself is called a spindle; however, in shop-floor practice, the word often is used to refer to the entire rotary unit, including not only the shaft itself, but also its bearings and anything attached to it (chuck, etc.).

A machine tool may have several spindles, such as the headstock and tailstock spindles on a bench lathe. The main spindle is usually the biggest one. Some machine tools that specialize in high-volume mass production have a group of 4, 6, or even more main spindles. These are called multi-spindle machines. For example, gang drills and many screw machines are multi-spindle machines. Although a bench lathe has more than one spindle (counting the tailstock), it is not called a multi-spindle machine; it has one main spindle.

Examples of spindles include:

- On a lathe, the spindle is the heart of the headstock.
- In rotating-cutter machinery, the spindle is the part on which shaped milling cutters are mounted for cutting features.
- Similarly, in rotating-cutter machine tools (such as milling machines and drill presses), the spindle is the shaft to which the tool (such as a drill bit or milling cutter) is attached.
- Varieties of spindles include grinding spindles, electric spindles, machine tool spindles, low-speed spindles, and high-speed spindles.

Spindle drive

Electric motors are used for spindle drive. Direct current and alternating current are used to drive the electric motors. They provide the power to the spindle.

Slide drive

There are a number of very effective, responsive, and thoroughly proved hydraulic systems currently in use, but by far the most common power source is the electric motor.

CNC controller

CNC is the automation of machine tools that are operated by precisely programmed commands encoded on a storage medium. Most CNC controllers today, just like computers, play an integral part of the control.

4.2.3 Some new applications of CNC machine tools

With the development of technology, CNC machine tools have some new applications, such as DNC, FMS, and CIMS.

DNC

"DNC" is a common manufacturing term for networking CNC machine tools. It can be understood as direct numerical control. It can also be termed as "distributed numerical control." On some CNC machine controllers, the available memory is too small to contain the machining program, so in this case the program is stored in a separate computer and sent directly to the machine, one block at a time. If the computer is connected to a number of machines, it can distribute programs to different machines as required. Usually, the manufacturer of the control provides suitable DNC software. However, if this provision is not possible, some software companies provide DNC applications that fulfill the purpose. DNC networking or DNC communication is always required when CAM programs are to run on some CNC machine tools.

From this point of view, the system contains a number of CNC machines rather than a single one. All the machines are linked to a computer which sends the information to the individual machines as required. There are a number of different types of machines linked together. In this situation, DNC can be understood as distributed numerical control. The computer is programmed to be able to select the order of manufacture of the components. DNC is capable of being integrated into the flexible manufacturing systems.

With the development of wireless communications technology, Wireless DNC is also used in place of hard-wired versions.

FMS

A flexible manufacturing system (FMS) is a manufacturing system in which there is some amount of flexibility that allows the system to react in case of changes, whether predicted or unpredicted. FMS usually consists of robots, CNC machine tools, instrumentation devices, computers, sensors, and other stand-alone systems such as inspection machines. The use of robots in the production segment of manufacturing industries promises a variety of benefits ranging from high utilization to high volume of productivity. Each robotic cell or node will be located along a material handling system such as a conveyor or automatic guided vehicle (AGV). The production of each part or workpiece will require a different combination of manufacturing nodes. The movement of parts from one node to another is done through the material handling system. At the end of part processing, the finished parts will be routed to an automatic inspection node, and subsequently unloaded from the flexible manufacturing system.

CIMS

With the release of Dr. Joseph Harrington's book, the idea of computer integrated manufacturing (CIM) became prominent in the early 1970s. However, it was not until 1984 when CIM began to be developed and promoted by machine tool manufacturers and the Computer and Automated Systems Association and Society of Manufacturing

Engineers (CASA/SME). CIM is the integration of total manufacturing enterprise by using integrated systems and data communication coupled with new managerial philosophies that improve organizational and personnel efficiency.

The term “CIM” is both a method of manufacturing and the name of a computer-automated system in which individual engineering, production, marketing, and support functions of a manufacturing enterprise are organized. In a CIM system, functional areas such as design, analysis, planning, purchasing, cost accounting, inventory control, and distribution are linked through the computers with factory floor functions such as materials handling and management, providing direct control and monitoring of all the operations.

4.2.4 Typical CNC machine tools

CNC machine tools can be categorized as CNC mills, CNC lathes, CNC machine centers, and CNC turning centers. The CNC machining centers are capable of multiple operations such as turning, milling, drilling, boring, and tapping operations, and can work on more than one face of the component.

CNC mills

CNC mills use computer controller to cut different materials. They are able to translate programs consisting of a specific number and letters to move the spindle to various locations and depths. Many mills use G-code, which is a standardized programming language that many CNC machines understand, while others use proprietary languages created by their manufacturers. These proprietary languages while often simpler than G-code are not transferable to other machines. CNC mills have many functions including face milling, shoulder milling, tapping, drilling, and some even offer turning. Standard CNC mills have three axes (X, Y, and Z) but others may also have one or more rotational axes.

CNC lathes

CNC lathe varies with different manufacturers, but they all have some common elements. The turret holds the tool holders and indexes them as needed, the spindle holds the workpiece using chuck or collet, and there are slides that let the turret move in multiple axes simultaneously. The machines are often totally enclosed. They have advantages such as ease of setting, operation, repeatability, and accuracy.

Machining centers

Vertical machining centers (VMC)

A vertical machining center has the spindle in the vertical direction as shown in Figure 4.1. An automatic tool changer is mounted to the machine to allow tools to be loaded into the spindle without operator intervention.

Figure 4.1: Vertical machining center.

Some vertical machining centers have a rotary table or indexer mounted on the table to allow a part to be rotated during machining. If a true rotary table is attached, the rotation of the rotary table will provide additional dexterity. This rotary axis is commonly called the C axis, or possibly the B axis, depending on how the rotary axis is mounted on the machine.

The vertical machining center has an automatic tool changer equipped to allow tools to be automatically placed in the spindle.

Horizontal machining center

A horizontal machining center (HMC) has the spindle oriented in the horizontal direction as shown in Figure 4.2. It also has an automatic tool changer equipped to allow tools to be automatically placed in the spindle.

Double work tables are equipped in horizontal machining center. This means that if the workpiece is being machined on a work table, another setup can be made on the opposite of work table.

Turning centers and Swiss-style lathe

If the CNC lathe has automatic tool changer systems, it can be called turning center.

A Swiss-style CNC lathe is a specific design of lathe providing extreme accuracy. A Swiss-style lathe holds the workpiece with both a collet and a guide bushing. The collet sits behind the guide bushing, and the tools sit in front of the guide bushing, holding stationary on the Z axis. To cut lengthwise along the part, the tools will move in and the material itself will move back and forth along the Z axis. This allows all the

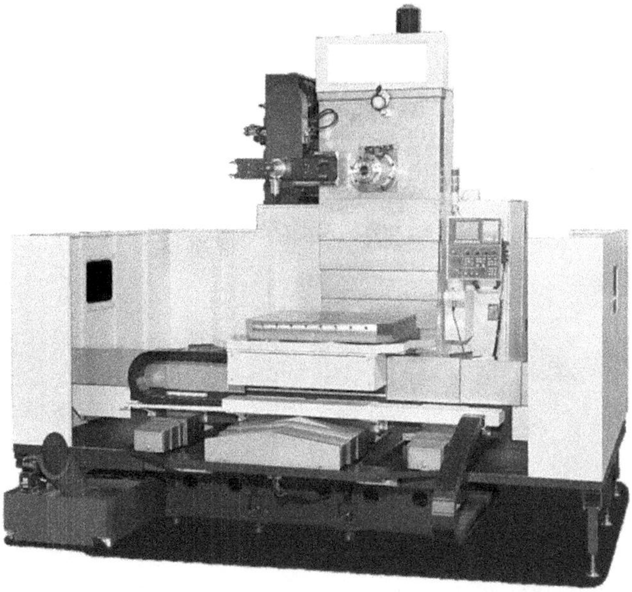

Figure 4.2: Horizontal machining center.

work to be done on the material near the guide bushing where it is more rigid, making them ideal for working on slender workpieces as the part is held firmly with little chance of deflection or vibration occurring.

Most CNC Swiss-style lathes today use one or two main spindles plus one or two back spindles (secondary spindles). The main spindle is used with the guide bushing for the main machining operations. The secondary spindle is located behind the part, aligned on the Z axis. In simple operation, it picks up the part as it is cut off, and accepts it for second operations, then ejects it into a bin, eliminating the need to have an operator manually change each part, as is often the case with standard CNC turning centers. This makes them very efficient, as these machines are capable of fast cycle times, producing simple parts in one cycle, in as little as 10–15 seconds. This makes them ideal for large production runs of small-diameter parts.

Additionally, as many Swiss lathes incorporate a secondary spindle, or "sub-spindle," they also incorporate "live tooling." Live tools are rotary cutting tools that are powered by a small motor independently of the spindle motor(s). Live tools increase the capability of the Swiss lathe. For instance, automatically producing a part with a hole drilled perpendicular to the main axis (the axis of rotation of the spindles) is very economical with live tooling, and similarly uneconomical if done as a secondary operation after machining by the Swiss lathe is complete. Generally, advanced CAD/CAM software uses live tools in addition to the main spindles so that most parts that can be drawn by a CAD system can actually be manufactured by the machines that the CAD/CAM software supports. Figure 4.3 shows a Swiss-style turning center.

Figure 4.3: Swiss-style turning center.

4.2.5 Tooling for CNC machine tools

Material for cutting tools

There are many types of material used to make cutting tools, ranging from high-speed steel to ceramics and diamond.

A cutting tool must have the following characteristics:

- Hardness: Hardness and strength of the cutting tool must be maintained at elevated temperatures. This is called as hot hardness.
- Toughness: Toughness of cutting tools is needed so that tools do not chip or fracture, especially during interrupted cutting operations.
- Wear resistance: Wear resistance means the attainment of acceptable tool life before tools need to be replaced.

High-speed steel

High-speed steel (HSS) is a subset of tool steels, commonly used in tool bits and cutting tools. It is often used in power-saw blades and drill bits. It is superior to the older high-carbon steel tools used extensively through the 1940s in that it can withstand higher temperatures without losing its hardness. This property allows HSS to cut faster than high-carbon steel, hence the name high-speed steel. At room temperature, in their generally recommended heat treatment, HSS grades generally display high hardness (above HRC60) and abrasion resistance (generally linked to tungsten and vanadium content often used in HSS) compared with common high-carbon steel tools.

Cemented carbide

The hardness of cemented carbide is almost equal to that of diamond. It derives this hardness from its main constituent, tungsten carbide. But the lack of toughness for cemented carbide presents problems. Carbides when coated can give superior properties.

Coating materials

Many types of coating materials are used for specific applications. The most common coating materials are titanium carbide, titanium nitride, ceramic coating, diamond coating, and titanium carbon nitride. In general, the coating process is accomplished by chemical vapor deposition (CVD). The substrate is placed in an environmentally controlled chamfer as a chemical vapor. The coating material is drawn to and deposited on the substrate by a magnetic field around the substrate. Physical vapor deposition (PVD) is another process except that the raw material that is going to be deposited starts out in solid form. Both processes are carried out under vacuum conditions.

In addition, diamond, cubic boron nitride (CBN), and so-called whisker-reinforced tools are used in some extreme cutting conditions.

Diamond

There are two types of diamond used as cutting tools: industrial grade natural diamonds and synthetic polycrystalline diamonds. Because diamonds are pure carbon, they have an affinity for the carbon of ferrous metals. Therefore, they can only be used on nonferrous metals.

CBN

CBN is similar to diamond in its polycrystalline structure and is also bonded to a carbide base. With the exception of titanium, or titanium-alloyed materials, CBN will work effectively as a cutting tool on most common work materials.

Whisker-reinforced tools

In order to further improve the performance and wear resistance of cutting tools to machine new materials and composites, whisker-reinforced composite cutting tool materials have been developed. Whisker-reinforced materials include silicon-nitride base tools and aluminum oxide base tools, reinforced with silicon carbide (SiC) whiskers. Such tools are effective in machining composites and nonferrous material, but are not suitable for machining irons and steels.

Tooling systems

Machining centers are multifunction machine tools. They use a wide variety of cutting tools such as drills, taps, reamers, end mills, and boring tools to perform various machining operations on a workpiece. A tooling system is an important feature of the modern machine tools. It helps to locate a range of cutting tools and securely held them in position.

Tool holders (flanges and collets)

The use of tool holders with standard tapers can be very helpful in keeping tooling costs to a minimum. Figure 4.4 shows some of the flanges and collets in a tooling system for machining centers.

Figure 4.4: Flanges and collets.

Three-piece system

A simple three-piece system typically includes a tool holder, an intermediate component, such as a boring bar or an insert-type milling cutter body, and an index insert cutter.

Automatic tool changer system

Automatic tool changer and tool holders are multipurpose devices that are designed to be easily manipulated by the tool changing mechanism, to ensure repeatability of a tool, i.e., place the tool in the spindle such that the tool's relation to the work is repeated within tolerance every time the tool is used, and to provide fast and easy off-line tool assembly.

Turret head

This type of system is found on older NC drilling machines. The tools are stored in the spindles of a device called a turret head. When a tool is called by the program, the turret rotates (indexed) it into position. The tool can be used immediately without having to be inserted into a spindle. Thus, turret head designs provide for very fast tool changes. The main disadvantage of turret head changers is the limit on the number of tool spindles that can be used.

Carousel storage

Systems of this type are usually found in vertical machining centers. Tools are stored in a coded drum, called a carousel.

Storage matrix magazine

Storage matrix magazine has been popular in machining centers since the early 1970s. This type of system permits an operator to load many tools in a relatively small space. A chain is located on the side or the top of the CNC machine tool. These positions enable tools to be stored away from the spindle and workpiece. This will ensure a minimum of chip interference with the storage mechanism and a maximum of tool protection.

When the system gets a programmed tool change command, the system advances the proper tool via the chain mechanism. The pivot arm of the tool change robot rotates and picks up both the new tool in the magazine and the old tool in the spindle. The magazine then advances to the space where the old tool is to be stored. The arm executes a rotation again and inserts the new tool into the spindle and the old one into the magazine. A final rotation returns the arm back to its parked position.

In order to recognize the tool, two methods of tool identification are currently in use. One is the bar code designation. The code is imprinted and fastened to the tool. When the program calls for a specific tool, the controller looks for a particular tool code, not a specific location. Another tool identification system uses a computer microchip, which is part of the tool or tool holder. The microchip contains the tool identification number and information related to the parameters of the tool. A special sensor reads the data and transfers it to the machine controller. A horizontal storage matrix magazine with pivot insertion tool changer is shown in Figure 4.5.

Figure 4.5: A horizontal storage matrix magazine with pivot insertion tool changer.

In Toyoda company in the United States, the matrix magazine is an auxiliary tool storage unit that increases the machine's standard capacity to up to 504 tools. Ideal for workpieces requiring many unique tools or multiple backup tools, this unit eliminates unnecessary trips to the tool center room.

http://www.toyodausa.com/machines/automation-solutions/matrix-magazines/matrix-magazine/

4.3 CNC programming

Up till now, part programming for CNC machine tools has been done normally by using ISO 6983. The CNC program is structured in blocks of data. A block of data consists of a sequence number word and one or more data words. Tab characters, which are optional for the tabulation of a printed copy of the data, may be inserted between words, but are usually ignored by the control systems. A data word always starts with an address character. The characters are usually presented in the following sequence:

- Preparatory function words G
- Dimension words. These words are arranged in the following sequence: X Y Z U V W P Q R A B C
- Interpolation or thread cutting lead words I J K
- Feed F
- Spindle speed S
- Tool T
- Miscellaneous M

Preparatory functions or G-codes are used to set the control for various machine movements, such as linear interpolation and rapid traverse. Linear interpolation (G01) means that the cutter moves on a controlled linear part (line). A G followed by a two-digit number determines the machine mode in that block or line. G codes or preparatory functions fall into two categories: modal and non-modal. Non-modal G-codes are those command codes that are only active in the block in which they are specified. Modal G codes are those command codes that will remain active until another code in the same group overrides it.

Interpolation is the typical way that CNC machining tools deal with tool path generations and execution for a given curve. Appropriate G codes are needed to be used to define the functional nature of the curve. For example, in linear interpolation, it works by effectively drawing a straight line between two neighboring samples and returning the appropriate point along that line. So G01 is used for linear interpolation, G02 and G03 are for circular interpolation, and G06 is for parabolic interpolation.

The coordinate system on a CNC machine is the basis of program. For a CNC machining center, the Z axis is usually first axis to be identified. It is identified by reference to a spindle. The X axis is horizontal and parallel to the work-holding surface.

The Y axis should be selected to complete with the X and Z a right-hand Cartesian coordinate system. A\B\C define rotary motions about axes parallel to X\Y\Z, respectively.

Compensations

Tool length compensation (G43 G44 G49 G48)

The CNC operator will not know for certain whether the tool length compensation value is perfectly correct until the first workpiece is machined. But he can use tool length compensation function to compensate for the difference in length between the tool assumed during programming and the tool to be used for actual machining.

G43 is a function for positive tool length compensation. With G43, a value from the offset register (OFFSET) can be called up and added to tool length. To all following Z movements in the program, this value will be added, while G44 is for negative tool length compensation. With G44, a value from the offset register can be called up and subtracted from tool length. To all following Z movements in the program, this value will be subtracted from. G49 is used to cancel tool length compensation; either the positive (G43) or negative (G44) shift will be cancelled. G48 is used to recall and enable tool length compensation.

Cutter radius compensation (G41 G42 G40)

Cutter radius compensation allows the operator to use the part geometry exactly as from the print for programmed coordinates. Without using compensation, the operator must always know the cutter size and offset the programmed coordinates for the geometry by the amount of the radius. Using cutter compensation, any size cutter can be used as long as the offset amount is input accurately into the offset register.

G41 is the function for 1eft tool nose radius compensation, while G42 is the function for right tool nose radius compensation. G40 cancels tool nose radius compensation.

Spindle Control

The rotating speed of the spindle can be controlled with spindle control command precisely. It is determined by the address S, which is used to indicate with rpm. If 350rpm is desired, the word S350 is used.

Feed Rate

The motion rate for any machining operation can be controlled with an F word. Feed rate is the velocity at which the cutter is fed, that is, advanced against the workpiece. It is expressed in units of distance per revolution for turning and boring (typically *millimeters per revolution*). It can also be expressed for milling, but it is often expressed in units of distance per time for milling with considerations of how many teeth (or flutes) the cutter has.

Tool Changes

Most machining centers have automatic tool changers that allow tools to be loaded into the spindle automatically during the program cycle. The machining center uses a

T word to rotate the machine's tool magazine to the desired position. For example, T04 rotates the magazine to station number 04. An M06 command is used to actually make the tool change and exchanges the tools in the ready position with tool in the spindle.

Miscellaneous functions M codes

Miscellaneous functions or M codes perform miscellaneous machine functions such as tool change, coolant control, and spindle operation. An M code is a two- or three-digit numerical value preceded by a letter address. M codes, similar to G codes, can be modal or non-modal.

Spindle on

The spindle can be turned on in the direction of clockwise rotation with M03.

Coolant

The coolant can be turned on or off at any time from within a program. An M08 command turns coolant on and an M09 turns the coolant off.

End of program

M30 is considered the standard program-end code, and returns execution to the top of the program.

4.4 Automatic programming tool

APT is a high-level computer programming language used to generate instructions for numerically controlled machine tools. The first prototype of the APT system was developed at MIT in the 1950s. APT can be considered as a CAM system based on a special purpose language. It was created to simplify the task of calculating geometry points that a tool must traverse in space to cut the increasingly complex parts required in the aerospace industry. It is a direct result of the new CNC technology becoming available at that time. It is capable of calculating the movements of the CNC for the complex parts. APT has a large vocabulary of the general processor languages. Because CAD systems are unavailable at that time, APT has to be able to describe the design geometry in text form. Therefore, there are three types of data defined in the APT language: part definition, machining plan, and machine specification.

Part definition

Part definition contains APT codes for describing some simple geometry in a part to be machined. These codes are only used to describe some of the key references and geometry necessary for the purpose of generating the CL data. For example, in order to describe a hole to be drilled, only a machining reference point and the center point of the hole are needed. The part geometry is defined in the Cartesian coordinate system, so coordinates are often used in defining geometry. There are typical APT geometry vocabulary words and statements used to define point, line, plane, circle, pattern, and cylinder.

Machining plan

There are three prime operating categories of numerical control: point to point, straight cut, and contouring. Point to point refers to operations that require rapid movement to a point followed by, or after, a manufacturing operation at that point. Such operations include drilling and punching. Straight cut indicates motion along only a major axis, for example, sawing. Contouring refers to the complex, continuous removal of material in an application such as turbine blade machining and pocket machining. Some typical machining plan commands are:

```
FROM/
GOTO/
GODLTA/
```

Looping of a subroutine and macro is often used to stand for a given sequence of instructions.

```
DRILLHOLE=MACRO
......
......
TERMAC
CALL DRILLHOLE
```

Machine specifications

Because CNC machine tools are different from one another, it may be that the controller is of a preparatory type, the cutters on the machine tool are handled differently, and the spindles have different power rating. As a result, a so-called post-processing step is required so that an APT program can be used on a particular CNC machine tool. Some of the typical post-processor commands are:

```
MACHIN/
COOLNT/
FEDRAT/
SPINDL/
TURRET/
```

CL files (Cutter location)

The cutter location data file is an NC program written in the APT language. This file is meant to be neutral because it is not formatted for any particular machine tool. There is neither a standard CL file format, nor the CL file content. Traditional APT CL file is binary and not human readable without translating. Since the APT language can define both geometry and tool motion data, a CL file will also contain the geometry and motion data. Now, it is in ASCII format, and is human readable.

CAD/CAM systems now have taken over the job of manipulating almost any geometry possible for machining purpose. There is no good reason to have geometry described in an ASCII APT CL file. With today's CAD/CAM system, design data can be easily used to generate machining information, and APT starts to fade away.

4.5 CAD/CAM integration

CAD/CAM integration technology is used to realize automatic transmission and conversion of component information among CAD and CAM applications. To understand CAD/CAM integration, it makes sense to start with the steps that need to be integrated.

4.5.1 Functions of CAD/CAM system

CAD is widely used to define a component with geometry, surfaces, or solid models. CAM is used to develop NC programs. Engineering design and manufacturing uses CAD/CAM software for three distinctly different purposes.

Design modeling A mechanical designer uses CAD software to create a part. This model can be represented as engineering drawings or CAD data files.

Manufacturing modeling A manufacturing engineer or CNC programmer uses CAD/CAM software to create new part models from the original design to allow for manufacturability. This would include developing models of the part for different steps in multiprocess manufacturing. This is part functions of CAPP. The programmer also uses CAD/CAM software to design models of fixtures, mold cavities, mold cores, mold bases, and other tooling.

NC programming A manufacturing engineer or CNC programmer uses CAM software to select tools, methods, and procedures to machine the models defined in the manufacturing modeling section described above. Note that the user that performs manufacturing modeling is usually the same user that performs CNC programming.

For these different products to work well together, they need to possess a high level of integration. Integration refers to how different software functions work together. There are three different types of integration to consider:

Data integration

Data integration is the ability to share part models (common data files). This is the most important type of integration for CAD/CAM. An IGES surface file represents poor data integration, due to the amount of manual repair work frequently required for success. A Parasolid file sent from one Parasolid-based software program to another represents very good data integration, as the data comes through flawlessly. Two Parasolid-based software programs sharing one Parasolid data file are even better, as both model history and associative can be maintained.

In the beginning, there were only CAD systems. Engineers used CAD systems to draw pictures of parts. The first CAM systems helped an NC programmer or manufacturing engineer program from these drawings. This making of drawings, and programming parts from drawings, was time consuming and subject to a lot of human error.

Afterwards, integrated CAD/CAM meant buying the same brand CAD and CAM products. Many companies provide such products today. These companies include PTC (Pro/Engineer and Pro/Manufacturing), Unigraphics (NX), Dassault (CATIA), and SDRC (IDEAS). All provide high-sophistication, high-power, and high-cost solutions. These products typically provide data, interface, and application integration. Because of their cost and complexity, these products do not provide ideal solutions for everyone. In addition, once the customer picks the CAD product he likes best, he's kind of stuck with whatever CAM product they have. No mixing and matching of products is allowed.

Another key change in the CAD/CAM market is the advent of the third-party solid modeler. Solid modeler kernel companies like Spatial Technologies (ACIS) and EDS (Parasolid) began selling to solution providers who started focusing on the mid-range CAD market. In 1995, SolidWorks and Intergraph SolidEdge were introduced and a new era of solid modeling CAD began. New product announcements have become common, with all major CAD companies jumping into this new market. Dassault (CATIA) had announced their intent to field a Windows NT product and recently purchased SolidWorks. Not only does solid modeling technology fuel the rapid growth of new products, but it also provides a backbone for seamless data transport between compatible products.

CAM in the new CAD Market

CAM products are also moving into this solid modeling CAD world. Some CAM products have plotted the shortest possible path to a marketing claim of solid-based solution. In several cases, this path has led to an inside CAD application version of their product, where the product's CAM capabilities are actually placed within the CAD software. This, however, can lead to a very limited set of CAM capabilities, not nearly the full feature set of the stand-alone versions.

There is another alternative. Any CAD or CAM product based on the same modeler can exchange data, providing a high level of data integration between different brand of products, for the first time. Solid model standards include Spatial Technologies' ACIS (.sat files) and Parasolid (.x_t and .xmt files). These standards are becoming widely supported.

OLE is another Windows technology that is becoming popular. This technology allows a CAM product to directly ask a CAD system for model data, without saving or opening files, or the technical problems of file translations. It is another powerful tool for providing data and application integration.

Historically, all CAD/CAM solutions offering high levels of integration have achieved it by putting all functions in the same computer program. This is one reason why some people think that there are big advantages with same application integration for CAD/CAM. Today's technology provides flexible alternatives to the same old application approach. Solid models and OLE provide excellent integration between different applications.

Visual verification

Most CAD/CAM software integrated simulation and verification that enables the CNC programmer to validate tool paths and machine motion through digital simulation form within the programming session. Using advanced simulation technology and full 3D in-process representations of the part, coupled with gauge and collision check methods, they eliminate the need to tie up expensive production equipment to verify manufacturability.

Being able to simulate and verify a manufacturing process in the digital world can significantly improve the expectation that the task will execute as planned, with no errors, in the real environment. In the area of CNC machining, there are a number of areas that the programmer and manufacturing engineer would frequently wish to validate before running a job on the machine tool. This ranges from a detailed check of the proposed tool path in terms of complete and correct material removal, without collisions or gauging, to a wider check on the overall machine motion to ensure that limits of travel are not exceeded and that there are no collisions with any part of the machine or fixtures.

4.5.2 CAD/CAM systems

Here are some examples of CAD/CAM software applications:

Master CAM

Master CAM is a software developed by CNC Software, Inc., which was founded in Massachusetts in 1983. Master CAM is one of the oldest developers of PC-based CAD/CAM software. It started as a 2D CAM system with CAD tools that let engineers design virtual parts on a computer screen and also guided CNC machine tools in the manufacture of parts. Since then, MasterCAM has grown into the most widely used CAD/CAM package in the world. Master CAM is an NC programming software for 2-axis, 3-axis, simultaneous 4- and 5-axis milling, multi-axis lathe, mill-turn, wire EDM, and routers.

CimatronE

Cimatron is an Israel software company that produces CAD/CAM software for manufacturing, toolmaking, and CNC programming applications. It is a leading global provider of CAD/CAM software solutions to the manufacturing market. Its main software product, CimatronE, is used in over 40,000 installations worldwide. The company was founded in 1982 as MicroCAD, releasing its first software products Multicad and Multicam in 1984 for use by small- to medium-sized tool shops. In 1987, the company changed its name to Cimatron. In 1990, the company launched Cimatron IT, which it claimed was the world's first integrated CAD/CAM software. In 2008, Cimatron merged with US CNC machining software company Gibbs.

CimatronE is an integrated CAD/CAM solution for mold, die, and tool makers, and manufacturers of discrete parts, providing association across the manufacturing process from quoting, through design and delivery. The solution's products include mold design, electrode design, die design, 2.5–5-axis NC programming and 5-axis discrete part production.

GibbsCAM

GibbsCAM was originally developed by US CNC machining software company Gibbs. It is one of the powerful CNC machine programming software that provides seamless integration to any CAD package. GibbsCAM specializes in 2- through 5-axis milling, turning, mill/turning, multitask simultaneous machining, and wire-EDM. It also provides integrated manufacturing modeling, including 2D, 2.5D, 3D wireframe, and surface and solid modeling. GibbsCAM was merged with Cimatron in 2008.

NX CAM

NX CAM and CAM Express allow CNC programmers to maximize the value of their investments in the latest, most efficient, and most capable machine tools. NX CAM provides the full range of functions to address high-speed surface machining, multifunction mill-turning, and 5-axis machining. CAM Express provides powerful CNC programming with low total cost of ownership.

For the software packages of Pro/Engineer and CATIA, they also have the functional modules of CAM.

4.6 STEP-NC

STEP-NC is a new programming interface for modern CNC-control. It supports bidirectional information flow between the shop floor and the planning.

STEP is originally developed as a neutral data standard for CAD data. Based on the previous ten years of effort to the STEP , STEP-NC uses the modern geometric constructs in that standard to define device-independent tool paths and CAM-independent volume removal features. The ISO 10303 AP238 standard replaced the ISO 6983 G code standard with a modern associative language that connects the CAD design data used to determine the machining requirements for an operation with the CAM process data that solves those requirements.

STEP-NC allows manufacturing organizations to seamlessly share machining and measurement information between machines and over the internet. The emerging STEP-NC technology is yielding process improvements of reduction in machining time. The automated measurement and compensation enabled by STEP-NC promises to yield parts that meet higher accuracy requirements for less cost. The integrated simulation and verification enabled by STEP-NC promises to guarantee that every part will be made correctly and that production will stop whenever cuts are about to be made that do not meet design requirements.

Working Steps

STEP-NC changes the way that manufacturing is done by defining data as “working steps”: a library of specific operations that might be performed on a CNC machine tool. In other words, it breaks down every machining operation into the steps required to perform the operation. In the past, CNC machine tools had to be programmed using G and M codes (from ISO 6983), which were instructions that only told the machine what moves to make, without any semantic content that referred to the part being processed. STEP-NC will make G and M codes obsolete.

With the concept of “working steps” in place, the manufacturing process becomes streamlined workflow. Now, a machine tool can receive a file with STEP-NC data, know what it means, and proceed milling the piece without any more instructions. There will be no more programming the machine tool for each individual piece. Moreover, the benefit of the standard goes further. With a set of standard working steps in place, all manufacturers will be able to share information reliably and instantaneously. A STEP-NC converted CAD file that is completed in China can be sent over the internet to a machine shop in the United States and they can immediately start milling the part.

4.7 The 3D printing technology

From a 3D digital model, people can make a physical object, typically by laying down many successive thin layers of a material. This is another way to integrate CAD and CAM, often called as 3D printing or additive manufacturing.

3D printing or additive manufacturing is a process of making solid objects from a digital file. Opposing to the material removing, or subtractive manufacturing processes, the creation of a 3D printed object is achieved using additive processes. In an additive process, an object is made by laying down successive layers of material. Each of these layers can be seen as a thinly sliced horizontal cross section of the eventual object.

4.7.1 The principle of 3D Printing

The working principle of 3D printing is entirely based on the principle of CAD/CAM integration. First of all, a virtual object should be designed. This virtual design is made in a CAD file using a 3D modeling program or with the use of a 3D scanner to copy an existing object. This scanner makes a 3D digital copy of an object and puts it into a 3D modeling program. It is often called as reverse engineering as mentioned in Chapter 3.

To prepare the digital file created in a 3D modeling program for printing, the software slices the final model into hundreds or thousands of horizontal layers. When this prepared file is uploaded in the 3D printer, the printer creates the object layer by layer. The 3D printer reads every slice (or 2D image) and proceeds to create the object blending each layer together with no sign of the layering visible, resulting in one 3D object.

4.7.2 Methods and technologies of 3D printing

Not all 3D printers use the same technology to print their objects. There are several methods to do it and all those available are additive. Some methods use melting or softening material to produce the layers. Selective laser sintering (SLS) and fused deposition modeling (FDM) are the most common technologies that use this way of printing. Another method of printing is to lay liquid materials that are cured with different technologies. The most common technology using this method is called stereo lithography (SL).

Stereo Lithography

Stereo lithography is a form of additive manufacturing technology used for creating models, prototypes, patterns, and production parts in a layer-by-layer fashion using photo-polymerization. In this process, light causes chains of molecules to link together, forming polymers. It works by focusing an ultraviolet (UV) laser on to a vat of photopolymer resin. With the help of CAD/CAM software, the UV laser is used to draw a preprogrammed design or shape on to the surface of the photopolymer vat. Because photopolymers are photosensitive under ultraviolet light, the resin is solidified and forms a single layer of the desired 3D object. This process is repeated for each layer of the design until the 3D object is complete.

It is also possible to print objects "bottom up" by using a vat with a flexible, transparent bottom, and focusing the UV laser upward through the bottom of the vat. An advantage of this bottom-up mode is that the build volume can be much bigger than the vat itself, and only enough photopolymer is needed to keep the bottom of the build vat continuously full of photopolymer. Often a depth of only a centimeter or two of liquid photopolymer is adequate to print thin structures 10 centimeters tall. Thus, bottom-up printing exposes far less photopolymer resin to atmospheric oxygen and room light, which can degrade or prematurely harden the photopolymer and so decreases the waste of the expensive photopolymer.

SL requires the use of supporting structures that attach to the elevator platform to prevent deformation due to gravity and to hold cross sections in place in order to resist lateral pressure from the resin-filled blade or retain newly created sections during the "vat rocking" of bottom-up printing. Supports are created automatically during the preparation of 3D CAD models and can also be made manually. With more expensive stereo lithography models, these supports must be removed from the finished product manually.

Selective Laser Sintering

Selective laser sintering (SLS) is another additive manufacturing technique that uses a laser as the power source to sinter powdered material (typically metal), aiming the laser automatically at points in space defined by a 3D model, binding the material together to create a solid structure. SLS is a technology that has been used mainly for rapid prototyping and for low-volume production of component parts.

As an additive manufacturing layer technology, SLS involves the use of a high-power laser (for example, a carbon dioxide laser) to fuse small particles of plastic, metal, ceramic, or glass powders into a mass that has a desired 3D shape. The laser selectively fuses powdered material by scanning cross sections generated from a 3D digital description of the part from a CAD file on the surface of a powder bed. After each cross section is scanned, the powder bed is lowered by one layer thickness, a new layer of material is applied on top, and the process is repeated until the part is completed.

Because finished part density depends on peak laser power, rather than laser duration, a SLS machine typically uses a pulsed laser. The SLS machine preheats the bulk powder material in the powder bed somewhat below its melting point, to make it easier for the laser to raise the temperature of the selected regions to attain the melting point.

Fused deposition modeling

Fused deposition modeling (FDM) is another technique used for 3D printing. FDM works on an "additive" principle by laying down material in layers; a plastic filament or metal wire is unwound from a coil and supplies material to produce a part.

FDM begins with the use of a software that processes an STL file (stereolithography file format), mathematically slicing and orienting the model for the build process. The machine may dispense multiple materials to achieve different goals. The part is produced by extruding small flattened strings of molten material to form layers as the material hardens immediately after extrusion from the nozzle.

A plastic filament or metal wire is unwound from a coil and supplies material to an extrusion nozzle which can turn the flow on and off. There is typically a worm-drive that pushes the filament into the nozzle at a controlled rate. The nozzle is heated to melt the material. The thermoplastics are heated past their glass transition temperature and are then deposited by an extrusion head.

The nozzle can be moved in both horizontal and vertical directions by a numerically controlled mechanism. The nozzle follows a tool path controlled by a CAD/

CAM software package, and the part is built from the bottom up, one layer at a time. Stepper motors or servo motors are typically employed to move the extrusion head. The mechanism used is often an X-Y-Z rectilinear design, although other mechanical designs such as deltabot have also been employed.

Although as a printing technology FDM is very flexible, and it is capable of dealing with small overhangs by the support from lower layers, FDM generally has some restrictions on the slope of the overhang, and cannot produce unsupported stalactites.

Many materials are available, such as acrylonitrile butadiene styrene (ABS), polylactic acid (PLA), polycarbonate (PC), polyamide (PA), polystyrene (PS), and rubber, with different trade-offs between strength and temperature properties. In addition, even the color of a given thermoplastic material may affect the strength of the printed object.

4.7.3 Industrial and personal 3D printing applications

In the past 10 years, the term "3D printing" has become well-known and the technology has reached a broader public. Manufacturers often used the 3D printers in their design process to create prototypes for traditional manufacturing and research purposes. Using 3D printers for these purposes is called rapid prototyping. 3D printing is also used for rapid manufacturing. Rapid manufacturing is a new method of manufacturing where companies are using 3D printers for short-run custom manufacturing. In this way of manufacturing, the printed objects are not prototypes, but the actual end user product.

Personal 3D printing is mainly for hobbyists and enthusiasts and really started growing fast. Because of rapid development within this new market, printers are getting cheaper and cheaper. For about 2,000 yuans (RMB), people have been able to buy a printing kit and put together their own personal 3D printer, complete with any customizations they are capable of making. What really speeds the development is the open source idea. This rapid development of open source 3D printers is gaining interest in both the developed as well as the developing world. This technology can assist in sustainable development as it can be easily and economically obtained from readily available resources by local communities to meet their needs.

However, the real integration of the newer additive technologies into commercial production is essentially a matter of complementing subtractive methods rather than displacing them entirely. It is predicted by some additive manufacturing advocates that this technological development will change the nature of commerce, because end users will be able to do much of their own manufacturing rather than engaging in trade to buy products from other people and corporations.

4.8 Summary

In this chapter, the general point of view of CAM has been described. The narrow point of view of CAM is also focused on CNC programming. Different kinds of CNC machine tools are the devices by which the CAM functions are carried out. There are many approaches for CNC programming, from the simple manual programming edit, to the APT, and to the integrated CAD/CAM software. STEP-NC is a new programming interface for modern CNC-controls. It supports bidirectional information flow between the shop floor and the planning. Finally, the CAM functions are further used in 3D printing technology.

Questions

1. What is computer aided manufacturing?
2. What is the benefits of CAM?
3. What is computer numerical control?
4. What does a CNC machine tool consist of?
5. What are the basics of CNC programming?
6. What is APT?
7. Give a brief introduction to STEP-NC.
8. What are the methods and technologies of 3D printing?

Tasks

1. Design a simple part with a blind pocket. Determine the cutting strategy and output the NC program. Illustrate the machine process by simulation approach.
2. Import a simple part file from CAD software, and use 3D printing method to produce the part.

References

Gao J, Zheng N, Gindy., Extraction of machining features for CAD/CAM integration, International Journal of Advanced Manufacturing Technology, 2004, 24: 573–581.

http://www.step-nc.org/index.htm

http://www.steptools.com/library/stepnc/

http://www.upcscavenger.com/wiki/Computer-aided%20manufacturing/#page=wiki

5 Integration of CAD/CAPP/CAM

Questions before you read

1. What are the automation islands?
2. What is the shortcoming of the automation islands?
3. How can the CAD, CAPP, and CAM systems be integrated?
4. What are the neutral file formats?
5. Do you know STEP or ISO 10303?
6. Do you know STEP-NC or ISO 14969?

The goals of this chapter are to seek a complete solution for integrating the CAD, CAPP, and CAM systems to understand the basic concept of integrating CAD/CAPP/CAM and to learn the available neutral file formats, such as DXF, IGES, PEDS, and STEP. STEP and STEP-NC are further introduced to be used in supporting data exchange and integrating CAD/CAPP/CAM.

5.1 Introduction

As discussed in earlier three chapters, the design process is automated with computer-aided design (CAD) systems, the process planning is automated with computer-aided process planning (CAPP) systems, and the manufacturing process is automated with computer-aided manufacturing (CAM) systems. But the technologies developed for CAD, CAPP, and CAM are localized within each of their domains, forming the so-called individual "the islands of automation." The gaps among CAD、CAPP、CAM systems are very large. The integration of CAD/CAPP/CAM is necessary to share the data and make information flow through these systems either forward control or feedback control. Here the symbols "、" and "/" have different meanings. The punctuation mark "、" means that these systems are separated alone. The previous generation of computer-aided systems had boundaries between system types, as well as the data formats. The alone system has little relationship to other systems. Nowadays, the forward slash symbol "/" embodied with a new meaning, that is, the systems are integrated. Different computer-based systems exchange the information and data between each other. Here the integration of CAD/CAPP/CAM is the most important issue to consider.

Today, users can select computer-based systems with a large number of choices of systems. The CAD interoperability issue poses a challenge to industries such as automotive, aerospace, shipbuilding, heavy equipment, and high-tech original equipment manufacturers and their suppliers. One uses a CAD system in-house, yet needs to deliver designs to, or receive from, another system. CAD is just a representation of the products geometry and topology. It is only the beginning of the product development.

https://doi.org/10.1515/9783110573091-005

CAPP relies on the product model data provided by a CAD system. It is a common practice to use design features in a CAD model and manufacturing features in a CAPP or CAM system. The process plan is usually further processed in a CAM system to obtain a set of machine control data.

More companies than ever before are involved in manufacturing various parts of their end products using different subcontractors, many of whom are often geographically diverse. It is needed to share information among vendors involved in multidisciplinary projects. Transfer of data is necessary so that one organization can develop a CAD model, while another performs analysis work on the same model, and at the same time, a third organization is responsible for manufacturing the product. Hence, transfer of data is necessary.

5.2 Product data exchange

CAD data formats are governed by the mathematical modeling kernel on which CAD systems are built upon. A modeling kernel is a collection of classes and components composed of mathematical functions that perform specific modeling tasks. Most of the kernels provide graphical and rendering support, including hidden-line, wire-frame, and drafting. Over the years, various kernels have been developed and adopted by different CAD systems. Most of CAD/CAM systems use popular ones through licensing, for example, ACIS by Spatial Technology Corporation, and Parasolid by UGS.

ACIS

The ACIS modeling kernel is a set of software algorithms used for creating solid modeling packages. Software developers get licenses of ACIS routines from the developer, Spatial Technology Corp., to simplify the task of writing new solid modelers. The key benefit of this approach is that models created using software based on ACIS should run unchanged with other brands of ACIS-based modelers. This eliminates the need to use initial graphic exchange specification (IGES) translators for transferring model data back and forth among applications. ACIS-based packages have become commercially available for CAD/CAM and FEA software package. Output files from ACIS have the suffix "*.SAT".

Parasolid

Parasolid is also a modeling kernel that can be licensed by other companies for use in their three-dimensional (3D) computer graphics software products. It is now owned by Siemens PLM software. Most Parasolid files can use Parasolid effectively; users need to have fundamental knowledge of CAGD, computational geometry, and topology. Parasolid modeling kernel provides 3D representation capabilities for NX, Solid Edge, and Team Center solution. Some are proprietary, such as CATIA and Pro/Engineer (Table 5.1).

Table 5.1: CAD/CAM systems and their build-in kernels.

CAD/CAM applications	ACIS	Parasolid	Proprietary
AutoCAD	*		
CATIA			*
IronCAD	*	*	
Creo (Pro/Engineer)			*
SolidWorks		*	
SIEMENS PLM – NX		*	
CAXA	*	*	

Dual kernel CAD systems

Dual kernel CAD system has two different kernels built into one system, which means two different kernels are built in a CAD system. For example, IronCAD, formally an ACIS-only system, has now been twinned with Parasolid to become a dual kernel system. It is the first dual kernel system in the world to deliver Dual Kernel Collaboration Technology. It simultaneously uses both kernels, switching back and forth when needed. The dual kernel CAD system can work on models developed under either kernel. The switch from one kernel to another happens in a nanosecond, usually invisible to the user.

Another dual kernel CAD system is CAXA, which is a product design system from China. It also used Parasolid and ACIS kernels. This kind of integration is efficient when ACIS and Parasolid are the data formats involved.

However, there are a large number of CAD kernels and CAD data formats in the market. It has proven to be extremely difficult to build such a system for various kernels, and the size of the software package would be the sum of both kernels. Furthermore, there are many other CAD kernels in the market. Thus, this approach only partially solves the problems.

Direct data translators

This option directly translates the modeling data stored in the database from one CAD system format to another, usually in one step. With this type of integration, a CAD system can import a Pro/Engineer part and rewrite it to another vendor format such as CATIA or AutoCAD format.

Direct translators provide a satisfactory solution when only a small number of systems are involved, but as this number increases the number of translator programs that need to be written become prohibitive. In general, if modeling data is to be transferred between all possible pairs of n CAD/CAM systems, then the total number of translators, N, that must be written is given by

$$N = n \times (n-1)$$

Two translators must be written for each pair of systems. Adding one system to the existing n system would require writing $2n$ additional translators. In Figure 5.1, eight additional translators (shown as dashed) are needed to accommodate the addition of System 5.

Neutral translators

A true type of common translator converts a proprietary CAD data format into a neutral data format and vice versa. This neutral data format may be of an international or industry accepted data format. There are a few popular industry standards.

As for neutral translators, they do not suffer from the increasing number of programs that have to be written as in the case of direct ones. For the case of n systems, the total number of neutral translators that must be written is given by

$$N = 2n$$

For $n = 3$, both give the same number of translators. For $n>3$, the required number of translators based on the neutral format philosophy is less than those based on direct conversion.

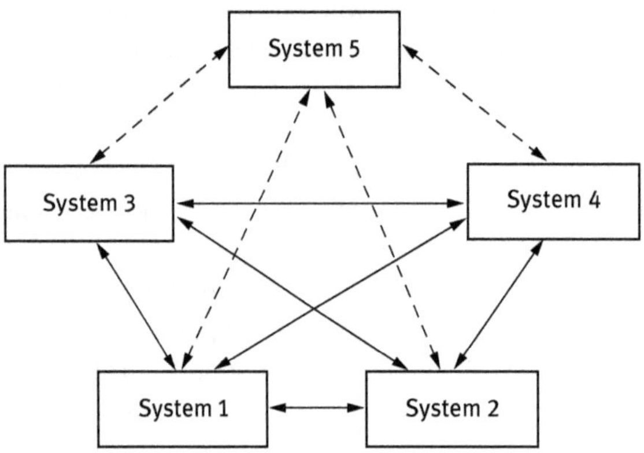

Figure 5.1: Direct data translators.

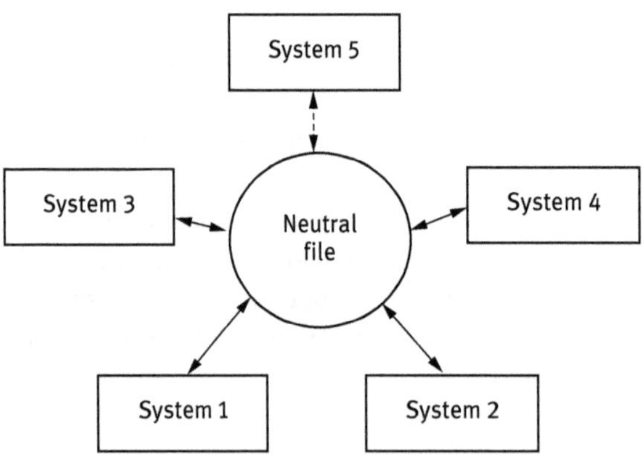

Figure 5.2: Neutral file translators.

Add one system (System 5 in Figure 5.2) to *n* existing systems would only require writing two additional translators (shown as dashed) regardless of the value of *n*. Moreover, neutral translator provides stable communication between CAD/CAM systems, protects against system obsolescence, and eliminates dependent on a single system supplier. Each type of translator has its advantages and disadvantages.

5.3 Some neutral data formats

In this section, some neutral data formats will be introduced. They are drawing eXchange format (DXF), IGES, product data exchange specification (PDES), and standard for the exchange of product (STEP) files.

5.3.1 Drawing eXchange Format

Originating in the late 1970s, drawing files created with microcomputer-based CAD software were saved with the .dwg extension. In the early 1980s, Autodesk® released AutoCAD®, which eventually became the most used CAD software in the world and which used Autodesk's undocumented and proprietary DWG™ file format (using the .dwg extension). DWG is the native and proprietary file format for AutoCAD® and a trademark of Autodesk, Inc.

DXF is a file extension for a graphic image format typically used with AutoCAD software. DXF stands for drawing exchange format. Since its initial release in 1982, there have been many changes to the DXF file format specifications. Depending on the software creating the DXF file, it can either be in an ASCII or a binary format.

With the DXF file format, Autodesk has got a solution for enabling data interoperability between AutoCAD and other programs. Solids and regions cannot be easily documented using DXF files.

DXF file structure

Essentially a DXF file is composed of pairs of codes and associated values. Each data element in the file is preceded by an integer number that is called a group code. The group codes indicate the type of value that follows. Using these group code and value pairs, a DXF file is organized into sections, composed of records, which are composed of a group code and a data item. Each group code and value is on its own line in the DXF file. Each section starts with a group code 0, followed by the string, SECTION. This is followed by a group code 2 and a string indicating the name of the section (e.g., HEADER). A section ends with a 0, followed by the string ENDSEC.

A DXF file consists of up to eight sections: header, classes, tables, blocks, entities, objects, thumb nail image, and an end-of-file mark. The overall organization of a DXF file is as follows:

HEADER
CLASSES
TABLES
BLOCKS
ENTITIES
OBJECTS
THUMB NAIL IMAGE
END OF FILE

The **HEADER** section contains general information about the drawing and variables that represent the state of AutoCAD's internal settings, such as the AutoCAD database version number and a number of system variables. Each parameter has a variable name and an associated value. For example, the AutoCAD version variable "$ACADVER" is set to "AC1015" in a DXF file saved by AutoCAD Release 2000. Other variables set the units used to measure angles, defaults for chamfering, offsets, and scalings.

The **CLASSES** section holds the information for application-defined classes, whose instances appear in the BLOCKS, ENTITIES, and OBJECTS sections of the database. A class definition is permanently fixed in class hierarchy.

The **TABLES** section contains several lists of information used in the rest of the drawing, such as the list of line types, layer names, fonts, and preset views of the drawing. Here it contains definitions of named items for the following symbol tables.

APPID (application identification table)
BLOCK_RECORD (block reference table)
DIMSTYLE (dimension style table)
LAYER (layer table)
LTYPE (linetype table)
STYLE (text style table)
UCS (User Coordinate System table)
VIEW (view table)
VPORT (viewport configuration table)

The **BLOCKS** section contains predefined drawing elements that might be present in the drawing. It contains blocks definition entities describing the entities comprising each block in the drawing. For example, a block could define a standard door knob that is placed on every door in a drawing. Block definitions are referenced in the ENTITIES section with the INSERT command.

The **ENTITIES** section contains the actual object data of the drawing, the drawing entities, including any block references. This can include raw data such as LINE and ARC entities as well as INSERT commands that place a predefined block definition at a certain position in the drawing.

The **OBJECTS** section contains nongraphical objects of the drawing. All entities that are not part of the entities or symbol tables are stored in this section.

Examples of entities in the OBJECTS section are dictionaries that contain mline styles and groups.

THUMB NAIL IMAGE preview image for the DXF file. This section is optional.

END OF FILE is the mark of the end of the DXF data directive on the last line of the file.

It may be helpful to produce a DXF file from a small drawing, print it out, and refer to it while reading the information presented in this section. Figure 5.3 shows a part of the file structure of DXF opened with notepad.

If you use the Select Objects option of the SAVE and SAVE AS commands, the resulting DXF file contains only the ENTITIES section and the END OF FILE mark. The ENTITIES section contains only the objects you select for output. If you select an insert entity, the corresponding block definition is not included in the output file.

Group codes and the associated values define a specific aspect of an object or entity. The line immediately following the group code is the associated value. This value can be a string, an integer, or a floating-point value, such as the X coordinate of a point. The lines following the second line of the group, if any, are determined by the group definition and the data associated with the group.

Special group codes are used as file separators, such as marks for the beginning and end of sections, tables, and the end of the file itself.

Entities, objects, classes, tables and table entries, and file separators are introduced with a 0 group code that is followed by a name describing the group.

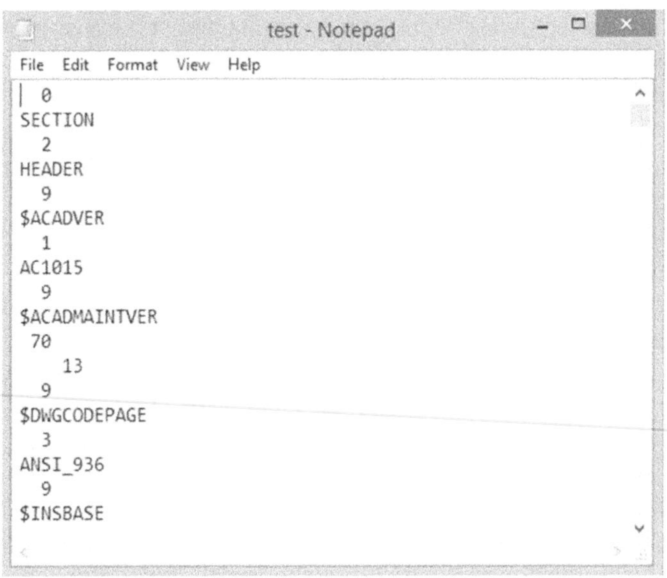

Figure 5.3: A part of the file structure of DXF opened with notepad.

The maximum DXF file string length is 256 characters. If the AutoCAD drawing contains strings that exceed this number, those strings are truncated during SAVE, SAVE AS, and WBLOCK. The functions of OPEN and INSERT will fail if the DXF file contains strings that exceed this number.

5.3.2 Initial graphic exchange specification

The IGES serves as a neutral data format to transfer the design to a dissimilar system. It is the first standard exchange format developed to address the concept of communicating product data among CAD/CAM systems. IGES is the ANSI standard Y14.26M. Translators, developed to the IGES Standard, are used to export a design into an IGES file for exchange and for importing the IGES file into the destination system. Using IGES, CAD designer can exchange product data models in the form of circuit diagrams, wireframe, free form surface, or solid modeling representations. Applications supported by IGES include traditional engineering drawings, models for analysis, and other manufacturing functions. Currently it supports solid modeling, including both B-rep and constructive solid geometry schemes.

IGES file format

An IGES file is composed of 80-character ASCII records, a record length derived from the punched card era. The file is divided into five sections: Start, Global, Directory Entry, Parameter Data, and Terminate, indicated by the characters S, G, D, P, or T in column 73.

The **Start** section is a human-readable introduction to the file. It is commonly described as a prologue to the IGES file.

The **Global** section contains information describing the preprocessor and information needed by postprocessors to handle the file. The first two global parameters define the parameter delimiter and record delimiter characters if the default values ("comma" and "semicolon," respectively) are not used.

The **Directory Entry** section has one record for each entity in the file. The Directory Entry record for each entity is fixed in size and contains 20 fields of eight characters each, in two consecutive 80-character lines. Values are right justified in each field. With the exception of the fields numbered 10, 16, 17, 18, and 20, all fields in this section shall be either integer or pointer data types. The purposes of the Directory Entry section are to provide an index for the file and to contain attribute information for each entity. The order of the Directory Entry records within the Directory Entry section is arbitrary.

The **Parameter Data** section contains the parameter data associated with each entity. Parameter data is free formatted, with the first field always containing the entity type number. Therefore, even though the Parameter Data section tables do not show it, the entity type number and a parameter delimiter precede Index 1 of each entity in the exchange file. Two groups of parameters are defined at the end of the specified parameters for each entity. The first group of parameters may contain pointers to any combination of one or more of the following entities: Associativity Instance Entity, General Note Entity, and Text Template Entity. The second group of parameters may contain pointers to one or more properties or attribute tables. Either group of parameters, or both, may be defaulted (i.e., empty).

In Terminate section, there is only one line in the Terminate section of the file. It is divided into ten fields of eight columns each. The Terminate section shall be the last sequenced line of the file.

Entities are categorized as geometry and nongeometry. Geometry defines physical shape, include points, curves, surfaces, solids, and relations, and nongeometry serves to enrich the model by providing a viewing perspective and providing annotation and dimensions. Two segments are provided to store the information, attributes for the geometric entity. They are directory segment and the parameter segment. The characteristics and geometric information for an entity is split between two sections: one in a two record, fixed-length format (Directory Entry), and the other in a multiple record, comma delimited format (Parameter Data), as can be seen in a more human-readable representation of the file. The directory entry provides an index and includes descriptive attributes about the data. They are organized in the fixed fields and consistent across all entities. The parameter data provide the specific entity definition and are variable in length and format. The connection between attributes and data segments is made with bidirectional pointers.

Figures 5.4 and 5.5 show an IGES example. This part model is created in the SolidWorks environment, saved as IGS file format.

IGES transfers both 2D and 3D finite elements for an FEA type of analysis, but lacks transferring solid objects.

5.3.3 Product data exchange specification

PDES was designed to completely define a product for all applications over the whole life cycle, from design, analysis, test, inspection, to following downstream applications, such as process planning, CAM, computer numerical control (CNC), and so on. Product data in PDES include geometry, topology, tolerances, relationships, attributes, and features. These data are necessary to a part or assembly for the purpose of design, analysis, manufacture, test, inspection, and maintenance, or even sale and recycle. PDES is designed to be full information for all downstream applications and to be directly interpretable by these applications. The main types of data in PDES include the following:

- Administrative and control data
- Geometry such as points, curves, and surfaces
- Topology such as vertices, loops, and faces
- Tolerances

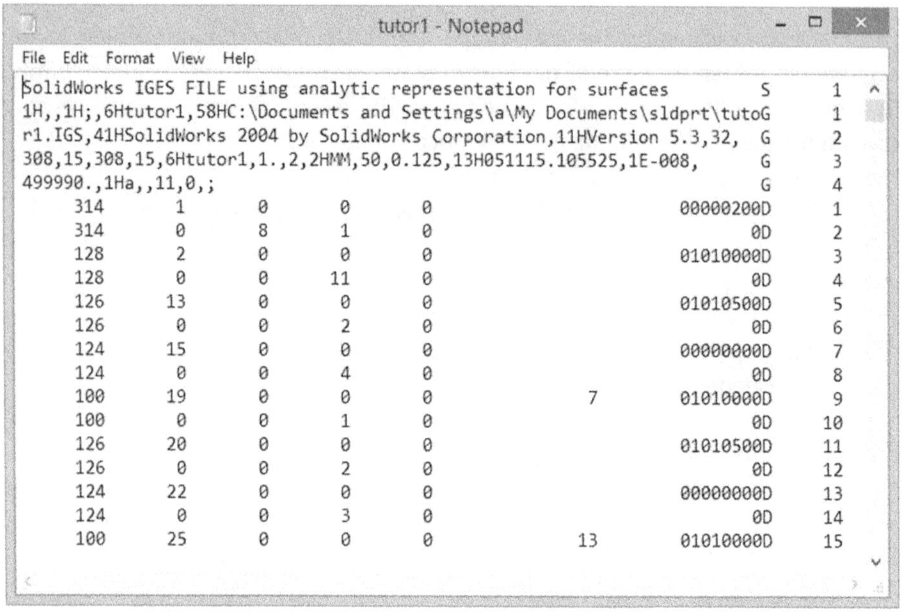

tutor1 - Notepad

File Edit Format View Help

```
SolidWorks IGES FILE using analytic representation for surfaces          S      1
1H,,1H;,6Htutor1,58HC:\Documents and Settings\a\My Documents\sldprt\tutoG      1
r1.IGS,41HSolidWorks 2004 by SolidWorks Corporation,11HVersion 5.3,32,   G      2
308,15,308,15,6Htutor1,1.,2,2HMM,50,0.125,13H051115.105525,1E-008,       G      3
499990.,1Ha,,11,0,;                                                      G      4
     314       1       0       0       0                        00000200D      1
     314       0       8       1       0                               0D      2
     128       2       0       0       0                        01010000D      3
     128       0       0      11       0                               0D      4
     126      13       0       0       0                        01010500D      5
     126       0       0       2       0                               0D      6
     124      15       0       0       0                        00000000D      7
     124       0       0       4       0                               0D      8
     100      19       0       0       0               7        01010000D      9
     100       0       0       1       0                               0D     10
     126      20       0       0       0                        01010500D     11
     126       0       0       2       0                               0D     12
     124      22       0       0       0                        00000000D     13
     124       0       0       3       0                               0D     14
     100      25       0       0       0              13        01010000D     15
```

Figure 5.4: The beginning part of an IGES file in notepad.

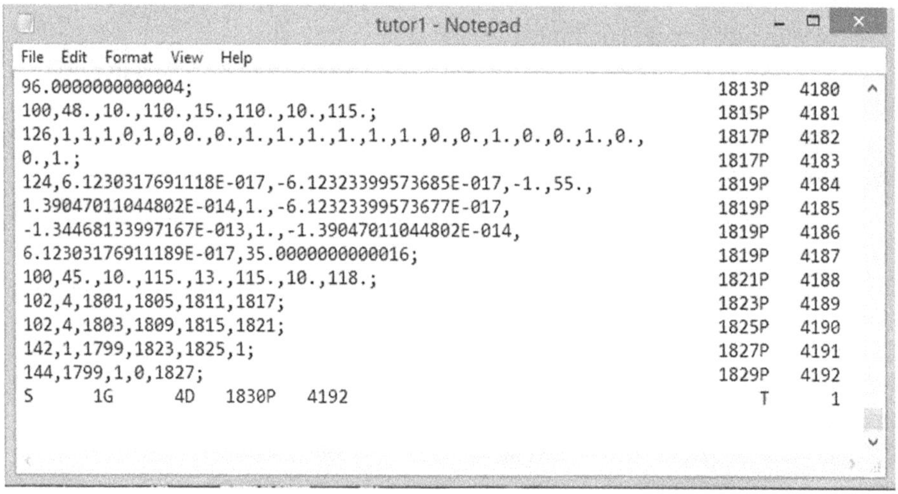

tutor1 - Notepad

File Edit Format View Help

```
96.0000000000004;                                                  1813P   4180
100,48.,10.,110.,15.,110.,10.,115.;                                1815P   4181
126,1,1,1,0,1,0,0.,0.,1.,1.,1.,1.,1.,1.,0.,0.,1.,0.,0.,1.,0.,      1817P   4182
0.,1.;                                                             1817P   4183
124,6.1230317691118E-017,-6.12323399573685E-017,-1.,55.,           1819P   4184
1.39047011044802E-014,1.,-6.12323399573677E-017,                   1819P   4185
-1.34468133997167E-013,1.,-1.39047011044802E-014,                  1819P   4186
6.12303176911189E-017,35.0000000000016;                            1819P   4187
100,45.,10.,115.,13.,115.,10.,118.;                                1821P   4188
102,4,1801,1805,1811,1817;                                         1823P   4189
102,4,1803,1809,1815,1821;                                         1825P   4190
142,1,1799,1823,1825,1;                                            1827P   4191
144,1799,1,0,1827;                                                 1829P   4192
S      1G      4D    1830P    4192                                     T      1
```

Figure 5.5: The end part of an IGES file in notepad.

- Form features
- Attributes such as surface finish
- Material properties
- Part assemblies

Here an example that shows how PDES describes a hole in the B-rep format.

Example: a hole in the B-Rep format

```
Hole: FLS1, FLS2
FLS1: FACE(LLS1, LLS2, SLS1)
FLS2: FACE(LLS2, SLS2)
SLS1: CYCLINDRICAL SURFACE(RAD1,AXIS2 PLACEMENT(CENT PT, DIRECTION)
P2 TSDO)
LLS1: ELS1
LLS2: ELS2
ELS1: EDGE(VTX1,VTX1,CLS1)
ELS2: EDGE(VTX2,VTX2,CLS2)
CLS1: CIRCLE(RAD1,AXIS2 PLACEMENT(CENT PT, DIRECTION)P1 TSD1)
CLS2: CIRCLE(RAD1,AXIS2 PLACEMENT(CENT PT, DIRECTION)P2 TSD1)
RAD1:0.500
P1:1.000,1.500,0.000
P2:1.000,1.500,-0.500
```

When a program reads the PDES file for the above hole, following parameters can be established.

There is one entity of type "Hole". Therefore, drawing code is given counter value H1;

Hole diameter = 2 × Hole radius = 2 × 0.500 = 1.000

Depth of hole is the Z distance between two points, P1 and P2: 0.500

Center point: P1(1.000,1.500,0.000)

PDES can be viewed as an expansion of IGES where organizational and technological data have been added. In fact, the later PDES contains IGES.

5.3.4 Standard for the exchange of product model data

STEP stands for standard for the exchange of product model data. It is an International Standard with code of 10303 for the computer-interpretable representation of product information and for the exchange of product data. The objective is to provide a neutral mechanism capable of describing products throughout their life cycle. This mechanism is suitable not only for neutral file exchange but also as a basis for implementing and sharing product databases, and as a basis for archiving. STEP is intended to support data exchange, data sharing, and data archiving. STEP continues most of the work in PDES. STEP is completely driven by industrial requirements.

An example of STEP

```
STEP;
HEADER;
```

```
FILE_NAME("EXAMPLE STEP FILE"),
"2017-10-03 T09:30:08",
("XUE Jianbin",
"Nanjing University of Aeronautics and Astronautics", "China"),
"STEP Version 1.0",
FILE_DESCRIPTION("This is a sample STEP file");
FILE_SCHEMA("GEOMETRY","TOPOLOGY");
ENDSEC;
DATA;
#1=PT(0.0,0.0,0.0);
#2=VX(#1);
ENDSEC;
ENDSTEP;
```

5.3.5 eXtensible Markup Language

XML stands for the eXtensible Markup Language, which has been used to represent EXPRESS schema and the data that is governed by the EXPRESS schema. The XML markup declaration enables XML representation of EXPRESS schemas.

5.4 Data exchange using STEP

The STEP standard was initially designed to offer a neutral data exchange method in replacement of IGES. However, the standard is much more than a neutral data format that translates geometrical data between CAD systems. The ultimate goal of STEP is to provide a complete computer-interpretable product data format, so that users can integrate business and technical data to support the whole product life cycle: design, analysis, manufacturing, sales, and customer service.

STEP is intended to support data exchange, data sharing, and data archiving. For data exchange, STEP defines the form of the product data that is to be transferred between a pair of applications. Each application holds its own copy of the product data in its own form. The data conforming to STEP are defined only for the purpose of exchange. STEP supports data sharing by providing access of and operation on a single copy of the same product data by more than one application. STEP is also suitable to support the interface to archive.

Components of STEP

The architectural components of STEP are reflected in the decomposition of the standard into several series of parts. Each part series contains one or more types of ISO 10303 parts. The STEP document architecture is shown in Figure 5.6.

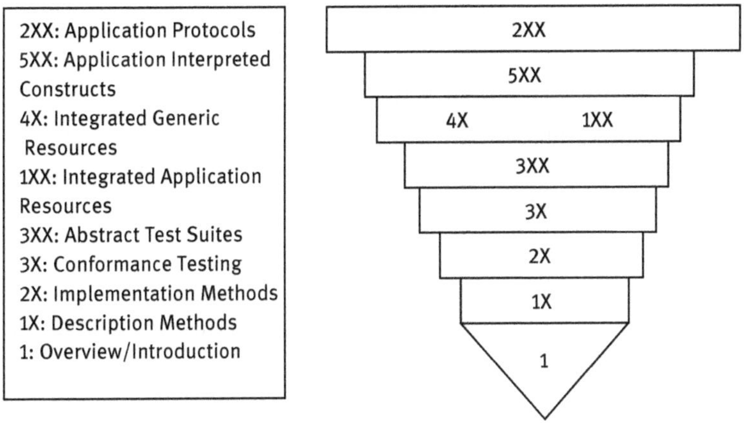

Figure 5.6: A document architecture of STEP.

Description methods

Description methods are common mechanisms for specifying the data constructs of STEP. They include the formal data specification language EXPRESS. Within a SCHEMA, various data types can be defined together with structural constraints and algorithmic rules. A graphical form of EXPRESS is called as EXPRESS-G. EXPRESS and EXPRESS-G are defined in ISO 10303-11. There are three main types of notations.

- Definition symbols denote simple data types, named data types, constructed data types, and schema declaration.
- Relationship symbols are different types of lines describing relationships that exist among the definitions.
- Supplementary text is used to further define a data entity or relationship, for example, an aggregation data type, constraints, and rules.

Implementation methods

Implementation methods are standard implementation techniques for the information structures specified by the only STEP data specifications intended for implementation, application protocols (APs). Each STEP implementation method defines the way in which the data constructs specified using STEP description methods are mapped to that implementation method.

This series includes the following:

- The physical file exchange structure (ISO 10303-21)
- The standard data access interface (ISO 10303-22)
- The language bindings (ISO 10303-23, 24, 27, 28)

Conformance testing

Conformance testing methodology and framework series provide an explicit framework for conformance and other types of testing as integral part of the standard.

This methodology describes how testing of implementations of various STEP parts is accomplished (ISO 10303-30).

An abstract test suite contains the set of abstract test cases necessary for conformance testing of an implementation of a STEP AP. Each abstract test case specifies input data to be provided to the implementation under test, along with information on how to assess the capabilities of the implementation.

Data specifications

Data specifications are the final major component of the STEP architecture. There are three primary types of data specifications: integrated resources, APs, and application interpreted constructs.

Three primary types of data specifications

- Integrated resources (ISO 10303-100)
- APs (ISO 10303-2XX)
- Application-interpreted constructs (ISO 10303-500)

Integrated resources represent concepts related to a particular application context that supports common requirement of many other product data applications. Examples of such resources include drawing sheet revision, drawing revision, and dimension callout. These may be used by any application that includes drawings. Integrated resources are standardized in the ISO 10303-100 series.

APs are the implementable data specification of STEP. APs include an EXPRESS information model that satisfies the specific product data needs of a given application context.

Application-interpreted constructs are data specifications that satisfy a specific product data need that arise in more than one application context. An application-interpreted construct specifies the data structures and semantics that are used to exchange product data common to two or more AP. Application-interpreted constructs are standardized in the ISO 10303-500 series of parts.

ISO 10303 is used for the computer-interpretable representation of product information and for the exchange of product data. There are some special APs used in data exchange, data transmission between different systems.

ISO 10303-203, 1994
Configuration controlled 3D designs of mechanical parts and assemblies,
ISO 10303-214, 1994
Core data for automotive mechanical design process

Currently, most of the commercial CAD systems can output STEP AP203 and/or STEP 214 files via STEP translators. By implementing STEP AP203 and STEP AP214 within CAD systems, data exchange barriers are only partially upheaved in a heterogeneous design environment. This is because both APs only document pure geometric information, leaving high-level data behind. Furthermore, data exchange problems between CAD/CAPP/CAM and CNC systems still remain unsolved.

5.5 AP 213 for CAPP

ISO 10303 – AP Part 213 is an AP for representing the data entities used in process plans for numerically controlled machined parts. AP213 is intended for archiving, exchanging, and sharing information between dissimilar CAPP systems and CNC programming systems. A numerical controlled machining process plan is a set of machining operations with a list of tools, setup instructions, and machining center specifications. The plan is used by numerical control (NC) programmers to create tool paths and NC programs that drive the tool motion to remove material from a workpiece. AP 213 specifies the data interface between process planning and CNC programming systems as opposed to the data necessary to perform the process planning functions.

5.6 STEP-NC (AP238 or ISO 14649) for CAM

The neutral data format such as STEP AP-203 and STEP AP-214 can work as the input data for CAM system. And the output of a CAM system is the ISO 6983 G-codes, which is a 60-year-old international standard, which still dominates the control systems of most CNC machine. This standard is out of data but still widely used. Thus, ISO 6983 has become an impediment for the modern collaborative manufacturing environment. Some of the technical limits and problems of G-code are summarized as follows:

1. G-code focuses on programming the path of the cutter location with respect to the machine axes, rather than the machining tasks with respect to the part.
2. The standard only supports one-way information flow from design to manufacturing without any feedback from shop floor.
3. Vendors usually extend the language out of the range of ISO 6983, so CNC are not exchangeable.
4. Users are limited in controlling the program execution. It is difficult to change the program in the shop floor.
5. The standard defines the syntax of a program statement, but in most cases the semantics is ambiguous.
6. ISO 6983 does not support the spline data, which makes it incapable of controlling five-axis milling.
7. CAD data are not utilized at machine tools. They have to be transferred into a set of low-level data by a machine-specific postprocessor.

These problems and limits make CNC machining a bottleneck in the current manufacturing environment. Modern CAD/CAM systems can handle complex programs with STEP-NC. The ultimate goal is to provide a complete product data format to support the whole life cycle. STEP has been extended to manufacturing data modeling and execution with an aim to fill the information gap between CAD/CAPP/CAM and CNC, that is, STEP-NC (AP238 or ISO 14649).

STEP-NC is a new language for CNC control. It makes CNC systems more interoperable, faster to program, and safer to operate. AP 238 of ISO 10303 specifies an AP for numerical controlled machining and associated processes, including the scope and information requirements defined by the ISO 14649 data model for numerical controllers, augmented with product geometry, geometric dimensioning and tolerancing, and product data management information. ISO 14649 defines a richer model for information transfer between CAD/CAM systems and CNC machines than that of the older ISO 6983 "G and M code" language. The ISO 6983 standard describes the path of the tool center point with respect to machine axes, while ISO 14649 describes machining processes with respect to a part, including part geometry, manufacturing features, sequencing of operations, associated process parameters, and tool requirements. It provides a complete, unambiguous model of the part and process.

This is a kind of task-oriented programming, as shown in Figure 5.7. AP-238 describes "what" as well as "how." It tells the machine tool to make this geometry from this stock by removing these features in this order with these tolerances, and tools that meet these requirements. It may even let the controller choose tool motion. While the old standard only described "how," it tells the machine tool to move tool to this location, move tool to that location, and so on for millions of commands.

AP-238 pulls together machining process, tool requirement, geometry, tolerances, and product data management (PDM), starting with process and tools described by ISO 14649 parts, AP238 adds STEP, geometric dimensions and tolerance (GD&T), PDM, geometry, and tolerances. These are common to all APs. These information are integrated so that features and all other data can be shared across APs.

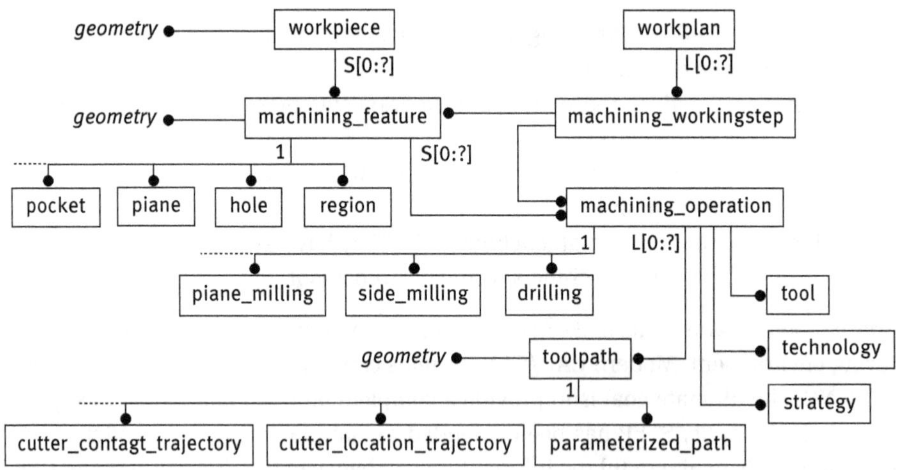

Figure 5.7: Task-oriented programming using AP-238.

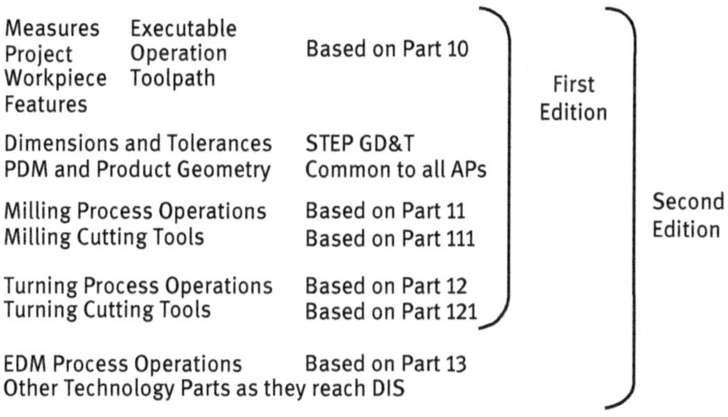

Figure 5.8: The structure of AP 238.

From the structure of AP-238 shown in Figure 5.8, it can be found that AP-238 describes the product or the workpiece in a PDM with product geometry, features, dimensions and tolerances, and measures. The workpiece is a STEP product description with all of the owner information, approvals, date and time, material, and material properties. Shape for workpiece uses advanced B-Rep, just like all of the other STEP APs. The information is produced by any CAD system that supports STEP.

There are lots of machining features should be considered in AP238. The 2.5D milling features include hole, pocket, slot, step, and so on. The transition features include edge round and chamfer. The replicate feature includes patterns of repeated feature. The region features include surfaces for free form milling. The turning features include outer-cylinder, knurl, and groove. Figure 5.9 shows some machining features.

For different machining processes, AP-238 describes the general process with project, executable operation, and toolpath.

The project is the starting place for machining program. It defines setup and main work plan. It is the heart of the STEP-NC model. Work plans contain a sequence of working steps. Each working step associates an operation with a feature somewhere on the workpiece. It is totally technology independent. Each operation describes what to do and what strategies and parameter settings to use. It has more information than G-codes, linked with geometric and technological information. Parameters can be grouped and shared, for example, tools, feeds, and so on. Intelligent CNC controllers can calculate tool movements for standard features.

The operations describe what a working step does to a feature, such as plane mill and center drill. It is the basis for all technology-specific extensions. It also defines the details like spindle, feed rate and other technology parameters, coolant and other machine functions, plunge strategies, tool paths, and so on.

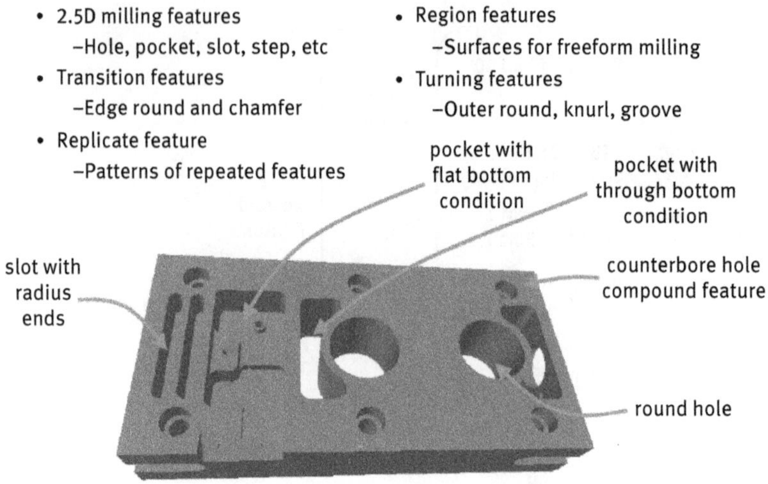

Figure 5.9: The machining features.

The toolpath is used for explicit control of the tool motion for an operation. It can use splines or other curves to describe motion and feed rate. It can describe path as machine-independent motion of tool center point or tool contact point. It can also describe path using old-style machine-dependent motion of individual axis.

AP-238 describes machining technology with drilling operations, milling process operations, and turning process operations. Drilling operations include drilling, boring, back boring, tapping, or threading. Tool moves only in the Z axis, and has many strategy parameters. Milling process operations include 2.5D machining operations and free form machining operations. About 2.5D machining operations are used for plane and side milling, roughing, and finishing. Tool can move XY or along Z axis, but not at the same time. It also has set of strategies. Freeform machining operations are special for sculptured surfaces. It controls 3, 4, and 5 axis sculptured motion. Turning operations include contouring, facing, grooving, threading, and knurling. It has roughing and finishing for each. Possible future operations work underway to describe operations for wire and sink electric discharge machining (EDM), contour cutting of wood and glass, and grinding. Efforts in various stages of maturity will be added to AP-238 once the information requirements are mature.

Every operation can have a tool associated with it. It only represents tool requirements, not a physical tool on a machine like the G-codes referenced tool #1, #2, and so on. With G-codes, human must make sure that #1 corresponds to the right thing. AP-238 can call out characteristic tool parameters for end mill, face mill, ball end mill, bull nose end mill, side mill, T-slot, tapered, dovetail, threading, and key seat mill. For drilling operations, AP-238 can call out tool for center drill, twist, tapered, or spade drills, combined drill and reamer or tap, and boring and reaming tools, tapered reamer, counter bore, countersink, backside counter bore and countersink, turning, grooving and knurling tools, and taps and threading tools.

In summary, AP-238 is a new language for CNC control. It is complete, unambiguous model of the part and process. It integrates machining process, tool requirement, geometry, tolerances, and PDM. It can enable million of dollars of savings using lean manufacturing by making CNC systems more interoperable, portable transmission control protocol (TCP) and no more posts. It is faster to program for task-oriented programming and safer to operate with full checking information.

5.7 Integration of CAD/CAPP/CAM

Data flow between CAD, CAPP, CAM, and CNC systems

By implementing STEP AP-203 and AP-214 within CAD systems, data exchange barriers are removed in the heterogeneous design environment. But data exchange problems between CAD, CAPP, and CAM remains unsolved. CAD systems are designed to describe the geometry of a product precisely. CAPP systems focus on using computer system to generate process plans, and control the manufacturing operations according to the geometric information presented in CAD models and the existing resource on the shop floor. The CAM systems generate a set of CNC programs G-code, or direct STEP-NC work plans that can be executed on a CNC machine. The neutral data format such STEP AP203 and AP214 only unify the input data for CAD system, and can be also used for CAPP and CAM systems. But on the output side of a CAM system, the 60-year-old ISO 6983 still dominates the control system of most CNC machines. ISO 6983 has become an impediment for the contemporary collaborative manufacturing environment. It makes CNC machining a bottleneck in the concurrent manufacturing environment. G-codes are now considered as hurdle in the way of increased productivity from state-of-the-art machine tools. Figure 5.10 shows the effect of an inadequate programming interface on the manufacturing information

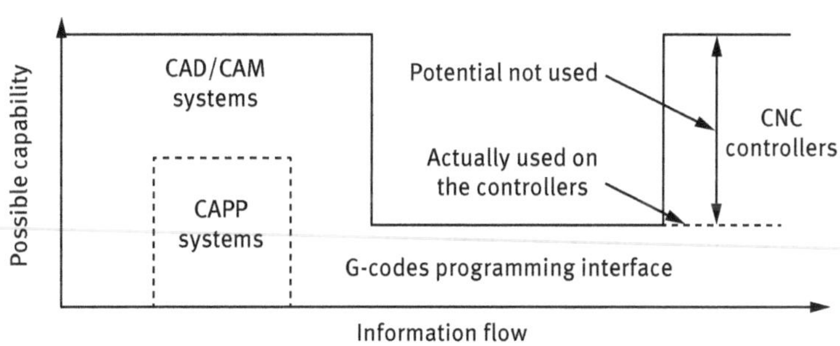

Figure 5.10: The information flow between CAD/CAPP/CAM/CNC.

chain. Modern CAD/CAM systems can handle complex programs. On the other side of the chain are the CNC controllers that are also able to perform complex work. Between them, the programming interface used the G-codes can only perform tasks that are limited in complexity and transfer little information regarding design and manufacturing data.

From the information flow between CAD/CAPP/CAM/CNC, the G-codes standard should be replaced by STEP and STEP-NC. Here in order to integrate CAD with CAPP, the ISO 10303 AP 203 and AP 214 are required to transfer product data, that is, CAD->CAPP (AP203, AP214). In order to integrate CAD with CAM, the ISO 10303 AP 203 and AP 214 are required to transfer product data, that is, CAD->CAM (AP203, AP214). While integrating the CAM and CNC, ISO 10303 AP 238 is required to transfer product data, that is, CAM->CNC (AP 238). The integration of CAD/CAPP/CAM can be realized with STEP and STEP-NC. ISO 10303 has also developed some new APs to integrate all processes during the whole product life cycle.

Integrate CAD with CAPP (AP 224)

CAD describes the geometry and topology of a product. CAPP generates plans and controls the manufacturing operations. AP 224 (mechanical product definition for process planning using machining features) was developed to bridge the gap between CAD and CAPP by providing machine part information that ensures the design information is 100% complete, accurate, computer interpretable, and reusable. Thus, automated process planning from digital product data is made possible. The AP 224 standard provides the mechanism to define the digital data that contains the information necessary to manufacture the required part. The information includes the following:

- CAD geometry and topology
- Machining feature information (e.g., hole, slot, groove, pocket, chamfer, and round)
- GD&T (dimension and geometric tolerance)
- Material and process properties
- Administrative information (approval, part name, and ID, delivery date and quantity)
- Assembly

AP 224 defines the implicit information about machining features; the dimension and geometric tolerance. AP 224 also defines part, process, material, and surface properties, even the part base shape, or the raw stock.

When several parts in an assembly have features in common, the machining features play an important role. For example, there are two parts that require four holes to be machined through both parts. The hole features and associated data such as tolerance and properties can be defined across both parts. A process plan would contain product data for both parts to be placed together on a machine, so that the hole could be machined through both parts at the same time. This is a special case in process planning, which can reduce the process time.

Integrate CAPP with CAM (AP 240)

AP 240 (ISO 10303-240, 2006) provides a standard way of defining process plan for machining a part. Integrating CAPP with CAM represents the process planning tasks. The process plan information for both numerical controlled and manually operated applications and associated product definition data are defined in AP 240. The following are the data defined in AP 240:

- Planning activity information
- Manufacture of a single piece part
- Technical data from upstream APs
- Work instructions for the tasks
- NC programming
- In-process inspection
- Shop-floor
- Production planning

Integrate CAM with CNC (AP 238)

After macroprocess planning, there comes the microprocess planning, which is closely related to a CNC machine. This has been done by AP 238 (ISO 10303-238, 2007) or ISO 14649 (ISO 14649-1, 2004), both are known as STEP-NC. AP 238 is the AIM of STEP-NC, while ISO 14649 is the ARM of STEP-NC. Being an ARM model, ISO 14649 provides a detailed analysis of the requirement of CNC applications. The specific objects and their relationships are accurately defined. The AIM model AP 238 maps the NC application requirement data into an AP. STEP-NC is used in a broad sense to refer to both or either of AP 238 and ISO 14649.

In order to support CAM/CNC integration, an open CNC architecture is developed based on the STEP-NC data model. The architecture intends to support bidirectional information flow in the design and manufacturing chain, to adopt the feature-based machining for CNC so that higher level information is made available at CNC machines, and to support a distributed process planning scenario.

5.8 Summary

The transfer of data between dissimilar CAD/CAM systems must embrace the complete description of a product stored in its database. Product data exchange is very important among different CAD/CAM systems, and all CAD/CAM vendors and many organizations have been collaborating to set exchange standard to make it manageable for users of CAD/CAPP/CAM system to communicate their product information effectively. The evolution of the exchange standards mimics the evolution of the CAD/CAM technology itself. Older standard such as IGES focus on geometric data, while newer standards such as STEP embrace the all types of product data. Other well-known standards such as ACIS and DXF have become popular due to their pervasive use in industry.

This chapter focuses on the data exchange between different CAD systems, and the data integration among different CAD/CAPP/CAM systems. The approaches of data exchange are introduced. Neutral data formats are the best solution to data exchange and system integration. Several file formats including DXF, IGES, STEP, and STEP-NC are described in detail. The STEP and STEP-NC are further analyzed. A conclusion can be drawn that the STEP and STEP-NC provide the best solution for the integration of CAD/CAPP/CAM.

Some of the STEP parts are listed below:

Part	Document Title
1	Overview and Fundamental Principles
11	EXPRESS Language
12	EXPRESS-I Language Reference Manual
21	Clear Text Encoding of the Exchange Structure
22	Standard Data Access Interface
23	SDAI C++ Language Binding
24	SDAI C Language Binding
25	SDAI FORTRAN Language Binding
31	Conformance Testing – General Concepts
32	Conformance Testing – Test Lab Requirements
33	Conformance Testing – Abstract Test Suites
34	Conformance Testing – Abstract Test Methods
41	Product Description and Support
42	Geometric and Topological Representation
43	Representation Structures
44	Product Structure Configuration
45	Materials
46	Visual Presentation
47	WShape Tolerances
48	Form Features
49	Process Structure and Properties
101	Draughting Resources
102	Ship Structures
103	Electrical/Electronics Connectivity
104	Finite Element Analysis
105	Kinematics

AP 201	Explicit Draughting
AP 202	Associative Draughting
AP 203	Configuration Controlled Design

AP 203e2	Configuration Controlled Design (second edition)
AP 204	Mechanical Design Using Boundary Representation
AP 205	Mechanical Design Using Surface Representation
AP 206	Mechanical Design Using Wireframe Representation
AP 207	Sheet Metal Dies and Blocks
AP 208	Life Cycle Product Change Process
AP 209	Composite and Metallic Structural Analysis and Related Design
AP 210	Electronic Printed Circuit Assembly Interconnect and Packaging Design
AP 211	Electronics Test Diagnostics and Remanufacture
AP 212	Electrotechnical Plants
AP 213	Numerical Control Process Plans for Machined Parts
AP 214	Core Data for Automotive Mechanical Design Processes
AP 215	Ship Arrangement
AP 216	Ship Moulded Form
AP 217	Ship Piping
AP 218	Ship Structures
AP 219	Dimensional Inspection Process Planning for CMMs
AP 220	Printed Circuit Assembly Manufacturing Planning
AP 221	Functional Data and Schematic Representation for Process Plans
AP 222	Design Engineering to Manufacturing for Composite Structures
AP 223	Exchange of Design and Manufacturing DPD for Composites
AP 224	Mechanical Product Definition for Process Planning
AP 225	Structural Building Elements Using Explicit Shape Rep
AP 226	Shipbuilding Mechanical Systems
AP 227	Plant Spatial Configuration
AP 232	Technical Data Packaging
AP 235	Engineering Properties
AP 236	Furniture Catalog and Interior Design
AP 238	STEP-NC Integrated CNC
AP 239	Product Life Cycle Support
AP 240	Macroprocess Planning
AP 242	Managed Model-Based 3D Engineering

ISO 10303-1:1994, Industrial automation systems – Product data representation and exchange – Part 1: Overview and fundamental principles

ISO 10303-11:1994, Industrial automation systems and integration – Product data representation and exchange – Part 11: Description methods: The EXPRESS language reference manual

ISO 10303-21:2002, Industrial automation systems and integration – Product data representation and exchange – Part 21: Implementation methods: Clear text encoding of the exchange structure

ISO 10303-28:2002, Industrial automation systems and integration – Product data representation and exchange – Part 28: Implementation methods: XML representations of EXPRESS schema

ISO 10303-41:2005, Industrial automation systems and integration – Product data and exchange – Part 41: Integrated generic resources: Fundamentals of product description and support

ISO 10303-42:2003, Industrial automation systems and integration – Product data and exchange – Part 42: Integrated generic resources: Geometric and topological representation

ISO 10303-43:2000, Industrial automation systems and integration – Product data and exchange – Part 43: Integrated generic resources: Representation structures

ISO 10303-44:2000, Industrial automation systems and integration – Product data and exchange – Part 44: Integrated generic resources: Product structure configuration

ISO 10303-45:1998, Industrial automation systems and integration – Product data and exchange – Part 45: Integrated generic resources: Materials

ISO 10303-47:1997, Industrial automation systems and integration – Product data and exchange – Part 47: Integrated generic resource: Shape variation tolerances

ISO 10303-49:1998, Industrial automation systems and integration – Product data and exchange – Part 49: Integrated generic resource: Process structure and properties

ISO 10303-501:2000, Industrial automation systems and integration – Product data representation and exchange – Part 501: Application interpreted construct: Edge-based wireframe

ISO 10303-502:2000, Industrial automation systems and integration – Product data representation and exchange – Part 502: Application interpreted construct: Shell-based wireframe

ISO 10303-507:2001, Industrial automation systems and integration – Product data representation and exchange – Part 507: Application interpreted construct: Geometrically bounded surface

ISO 10303-508:2001, Industrial automation systems and integration – Product data representation and exchange – Part 508: Application interpreted construct: Non-manifold surface

ISO 10303-509:2001, Industrial automation systems and integration – Product data representation and exchange – Part 509: Application interpreted construct: Manifold surface

ISO 10303-510:2000, Industrial automation systems and integration – Product data representation and exchange – Part 510: Application interpreted construct: Geometrically bounded wireframe

ISO 10303-511:2001,Industrial automation systems and integration – Product data representation and exchange – Part 511: Application interpreted construct: Topologically bounded surface

ISO 10303-512:1999, Industrial automation systems and integration – Product data representation and exchange – Part 512: Application interpreted construct: Faceted boundary representation

ISO 10303-514:1999, Industrial automation systems and integration – Product data representation and exchange – Part 514: Application interpreted construct: Advanced boundary representation

ISO 10303-519:2000, Industrial automation systems and integration – Product data representation and exchange – Part 519: Application interpreted construct: Geometric tolerances

ISO 10303-522:2006, Industrial automation systems and integration – Product data representation and exchange – Part 522: Application interpreted construct: Machining features

ISO 14649-1:2003, Industrial automation systems – Physical device control – Data model for computerized numerical controllers – Part 1: Overview and fundamental principles

ISO 14649-10:2004, Industrial automation systems – Physical device control – Data model for computerized numerical controllers – Part 10: General process data

ISO 14649-11:2004, Industrial automation systems – Physical device control – Data model for computerized numerical controllers – Part 11: Process data for milling

ISO 14649-12:2005, Industrial automation systems – Physical device control – Data model for computerized numerical controllers – Part 12: Process data for turning

ISO 14649-111: –1, Industrial automation systems – Physical device control – Data model for computerized numerical controllers – Part 111: Tools for milling

ISO 14649-121:2005, Industrial automation systems – Physical device control – Data model for computerized numerical controllers – Part 121: Tools for turning machines

Questions

1. What are the islands of automation?
2. What are the kernels of CAD systems?
3. How can you deal with the different data format?
4. What are the neutral translators?
5. What is the file structure of DXF?
6. What is the IGES file format?
7. What are the components of STEP?
8. What is the limitation of ISO 6983 G-code?
9. Describe the data flow between CAD, CAPP, and CAM systems.
10. What is the STEP-NC?
11. How can you integrate CAD with CAPP using STEP?

12. How can you integrate CAPP with CAM using STEP?
13. How can you integrate CAM with CNC using STEP?
14. Describe the paradigm of integration with STEP and STEP-NC.

Tasks

1. Design a simple part in a 3D CAD/CAM software, save as .igs, .stp, .stl file extension, use notepad to open them, and read these files focus on the sections of the file.
2. Draw a diagram to express how to integrate the CAD, CAPP, and CAM, or even CNC.

References

IGES/PDES Organization, Initial Graphics Exchange Specification: IGES 5.3, N. Charleston, SC: U.S. Product Data Association, 2006.

Xu X., Integrating advanced computer aided design, manufacturing, and numerical control, principles and implementations. Information Science Reference. 2007.

http://www.caxa.com/

http://www.ironcad.com/

6 Product data management

Questions before you read:
1. How can you deal with so many kinds of documents in engineering?
2. What is the general concept of product data management?
3. What are the main functions of PDM systems?
4. Do you know any well-known PDM system packages?
5. Do you know some latest developments in PDM/PLM technology?

The goal of this chapter is to understand in detail about product data management (PDM), including its concept, main functions, and the latest development in PDM/PLM technology. Several popular commercial PDM software packages are also introduced.

6.1 Introduction

Historically, different computer applications have been developed to handle certain parts of the product life cycle. CAD focuses on product design, CAE focuses on product analysis, CAPP focuses on process planning, and CAM focuses on product manufacturing. PDM handles the management of design and drafting files from conceptualization through detailed design. If the range covers the entire product life, then PDM is extended into product life cycle management (PLM). PLM are software-based computer systems used to manage the product data at the stages of design, analysis, process plan, manufacture, sale, and even discarding. The information, files and documents, and work processes are defined and managed in a certain mode to ensure consistency, reuse, and sharing in the entire product life cycle. They provide support for the activities of product teams and for techniques such as concurrent engineering and collaborative design.

Product data management derives from the concept of engineering data management (EDM), which is used for the effective management of product development data and the coordination of manufacturing-related processes. The term is used primarily in the field of CAD. As discussed in previous chapters, there are various file formats in different versions, which are often used by several users and in different departments. Product data management is very important for the product development.

There is a need for an effective way of managing product related information as well as the product development action flow, which captures actions that need to be done, have been done, and what other parts are affected. This leads the project management and workflow management to PDM and PLM.

PDM supports engineers when creating and managing product data for the entire life cycle of the product. It offers a complete solution for management of product data. It uses software to manage product data and process-related information in a single,

https://doi.org/10.1515/9783110573091-006

central system. This information includes CAD data, models, parts information, manufacturing instructions, requirements, notes, and documents.

6.2 Functions of PDM

The architecture of PDM system is divided into four levels: system supporting level, framework core level, level for function modules and developing tools, and the user interface level. Figure 6.1 shows the four levels of PDM systems.

The first level is the system supporting level, which includes operating systems, database systems, and network systems. Operating systems mean that PDM runs on Windows or UNIX system. Database systems provide space to store and retrieve the product data. Networks are the routes that the product data transfer.

The second level is the object management framework. Here, the framework still uses the existing relational database to manage the objects. It takes advantage of the mature data management technologies inherent in the relational database. It is easy to keep the data consistency, integrality, and concurrent access control. Adding more constraints on the object layer makes the system control more flexible. The tree structure of class makes the system more adaptive and open, and allows the users to add classes and functions to meet the requirements of their enterprises. But, the system has some problems, such as how to improve the efficiency of data retrieval, schema evolution, etc. Object management framework also needs to deal with the distributed

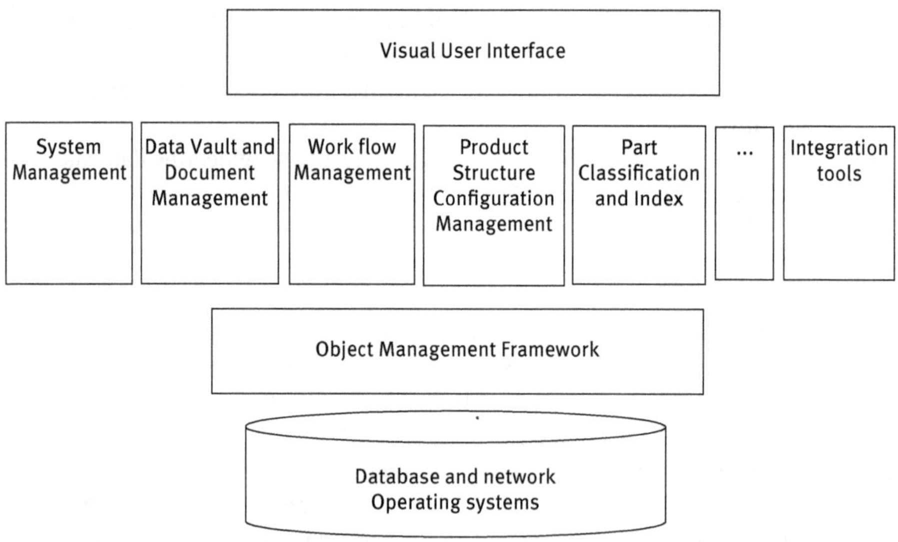

Figure 6.1: Architecture of common PDM systems.

applications. Both object-oriented programming and relational database management systems are extremely common in software today.

The third level is the level for function modules and developing tools. PDM has several groups of functionality. The function modules consist of the following five basic user functions supported by PDM systems:

1. Data vault and document management, which provides for storage and retrieval of product information;
2. Workflow and process management, which controls procedures for handling product data and provides a mechanism to drive a business with information;
3. Product structure management, which handles bills of material, product configurations, and associated versions and design variations;
4. Parts management, which provides information on standard components and facilitates reuse of designs;
5. Project management, which provides work breakdown structures and allows coordination between processes, resource scheduling, and project tracking.

Furthermore, there are some other utility functions that enhance PDM systems. Communication capabilities such as links to email provide for information transfer and events notification. Data transport function tracks data locations and moves data from one location or application to another. Data translation capability exchanges files in a proper format. Image services handle storage, access, and viewing of product information. Administration function controls and monitors system operation and security.

The PDM user can search company's data or information through his or her desktop computer. The actual searching and finding process is handled by the server, using the meta-database search engine. The files are stored in the managed files or data vault. The server finds the specific information and then transfers it back to the user's screen with proper format.

Data vault

Data vault is one of the main subsystems of PDM, which is used to manage the files and documents in a database. Data is stored in a central archive called a file vault, which can be regularly backed up and shared by the entire product team. To access files, a working folder is created on the local computer called a file vault view. Logging in to the file vault view, one can access files for which he has permissions. The files can be checked out to edit, so that no one else can make changes, though other users can still view and copy the files.

The data vault stores, organizes, controls, retrieves, and protects all product data. It supervises the access of those data by individual, project, or some other user-defined parameters. The data vault can contain the data itself or contain metadata, data about the location of the data. Users must go through the PDM system to get controlled data. It means that users don't have to know where data are actually located and ensure that they get the latest versions of data.

The types of data vault can be centralized, distributed, or virtualized. PDM system shall cover all product data. If data vault is centralized, all product data are stored in one computer. Other computers retrieve the data through network. If data vault is distributed, the product data are stored in individual computers dispersed in the company. Computers retrieve or store the data through network. In most cases, the data vault is virtualized. It is not known where the product data are stored. Anywhere is possible.

Operation of data vault includes check-in, check-out, and authorization. The function of check-in is used to check in the object to the PDM system to prevent other useless object stored in the PDM system. The check-out function prevents an object to be modified by different users at the same time. If a user is working on an object, no one else other than him can work on it. Other users can only have a view without editing access on the object at the same time.

Authorization is used to assign different rights to operate the objects in the PDM system.

File ownership, version control, check-in and check-out of files, revision management, and release status are all managed by the PDM system. Security and administrative functionality protect intellectual property rights through role management, project-based security, and associated access privileges.

Document Management

Beginning in the 1980s, a number of vendors began developing software systems to manage paper-based documents. These systems dealt with paper documents, which included not only printed and published documents, but also photographs, prints, etc. Later developers began to write a second type of system which could manage electronic documents, i.e., all those documents, or files, created on computers, and often stored on users' local file systems. The systems could manage any type of file format that could be stored on the network. The applications grew to encompass electronic documents, collaboration tools, security, workflow, and auditing capabilities. These systems enabled an organization to capture faxes and forms, to save copies of the documents as images, and to store the image files in the repository for security and quick retrieval. While many systems store documents in their native file format (Microsoft Word or Excel, PDF), some web-based document management systems are beginning to store content in the form of HTML. The HTML format allows for better application of search capabilities such as full-text searching and stemming searching.

Objects for document management include the design, process plan, production and manufacture, project management, and even sales management. These documents can be classified with the file formats. So the systems provide solutions for scanning, automation, classifying, indexing and archiving the text, data file, drawing, tables, and media files. These documents can be previewed and searched with navigator. There are some other functions of document management, such as classification and index, version management, security management, and compressing and decompressing files.

Workflow and process management

Workflow and process management capabilities enable both internal product teams and external partners to participate in the product life cycle. As a key element in both processes and configuration management, workflow processes automatically parcel the results of tasks to the next set of tasks in a series, and can alert the associated individuals that work is waiting or being accomplished. The work defined in a workflow system can be serial or concurrent; in either case, work can proceed based on the simple completion of a step or on conditional requirements. Repetitive processes are ideal for workflow operations. When given a graphical user interface, workflow lets users see where resources and deliverable sit in the overall engineering process. The workflow expressed in this way is considered as a Kanban in just in time (JIT) production environment.

A PDM system can be used to help establish, manage, and execute automated workflow-driven processes that reflect company-specific best practices for change planning (what-if analysis), change incorporation (execution), and change verification and communication.

Workflow and process management includes task management, workflow management, and task history management. Its functions include defining the workflow, defining the data flow, and assigning the users.

Product configuration management

Product configuration management considers the product structure management. This function focuses on creating and managing product configurations and bills of material (BOM). It defines the parts data relationships, and other product data throughout product development life cycle. Features include keeping and managing previous build versions, managing effectiveness, supervising several different configurations of the product simultaneously, managing engineering change orders, and providing different aspects of a product based on design discipline.

Engineering change management

Engineering change management is a general term for the engineering change request and engineering change order processes. Sometimes, engineering change management is called engineering change notification. Engineering change management is used to create, plan, review, approve, and implement product changes. New designs or products may be included in the engineering change order process or may go through a similar process called engineering creation order. Engineering change orders usually include changes or enhancements to existing products or process specifications. Changes are generally made to products to resolve quality and safety issues or to improve product performance. Engineering change management is a tool used throughout the organization to communicate changes to product data.

Bill of material

A PDM system provides the visibility necessary for managing and presenting a complete BOM. It facilitates the alignment and synchronization of all sources of BOM data, as well as all life cycle phases, including the as-designed, as-planned, as-built, and as-maintained states.

Life cycle visualization provides sharing and on-demand representations of the product and its underlying assemblies and parts, without the need for a CAD authoring tool or special technical knowledge.

Product structure can be expressed with file directories and documents. Version management is used to manage different version of product structure during development. Product configuration has certain rules of configuration. The bill of material can be used to express single product configuration. When the production has a serial of product, the serial products configuration should be established. And multi-views of product structure are used for different departments.

Accounting uses bills of material and routing to run a cost roll-up. From the cost roll-up, the cost of the product can be determined and so can the product price. The service department uses the parent/component relationships in bills of material to determine what parts need to be stocked for warranty and replacement parts.

Part coding systems and management

PDM-based coding systems are needed to classify the part family using certain coding systems. The parts are classified according to the attribute-based standard part, and the attribute-based similar parts. The coding system manages to establish relation between parts, documents, and part families. Part coding system and management maintains the classifying mechanism and the relationships between part families. This function can help engineers use CAPP systems.

Project Management

PDM-based project management includes real-time resource scheduling and project tracking, progress reporting, status and location of deliverable, and identifying the individuals working on deliverable. Project management is the least mature group of functionality within PDM, and is often provided by third-party applications. The project should be properly organized and controlled. The roles in the project should be defined to form several team groups. Each group of users has different right to operate the PDM system through authorization function. The project management also needs some useful tools such as Gantt chart, and CPM to help the execution of a project.

Customization and Integration functions

According to the users' requirement, some tailor, modification, and adding functions are made to the PDM system. Some links between PDM system and other systems are needed for the integration.

Other functions

The PDM system also has some other functions, such as scanning, reading and noting, etc.

6.3 PDM software vendors

The PDM software products are commercially available in the PDM markets. There are many vendors who offer their products to different customers. Some are directed at large organizations, others are PC-based and intended for small departments, still others are focused on specific industries; in addition, some are bundled with other applications, while others are offered independently. Some well-known examples of PDM software applications are discussed next.

SDRC Metaphase

In the 1990s, SDRC Metaphase system was one of the most popular PDM systems. Metaphase provides the foundation for the Four C's of e-PDM – key business process improvements companies must have to succeed in e-business: the capability to control information, connect people with information, facilitate collaboration among information users, and configure products based on fast and accurate information sharing. The Metaphase web-centric information infrastructure is the basis for its flexible and scalable solutions that are quickly implemented in companies of all sizes for immediate business return in the e-business global marketplace. It allows customers to identify and understand the impact of changes on product requirements, design, cost, scheduling, and quality, and to authorize modifications to product structures and definition information when changes occur.

Siemens UG Teamcenter

Teamcenter is the world's most widely used PLM software, more than just a PDM solution. At first, IMan (Information Management) system was a PDM system developed by EDS. Later, it was bought out by Siemens and renamed as Teamcenter. Teamcenter connects people throughout the product life cycle with a single source of product and process knowledge. Teamcenter Rapid Start is a pre-configured PDM solution that delivers the most common industry best practices and the expertise of Siemens PLM Software.

PTC Windchill

PTC Windchill is a PDM software product that offers comprehensive capabilities to help manufacturers manage their products through all phases of the product life cycle. Windchill manages all product content and business developments throughout the product and service life cycle with various modules that support product development. Built in an internet-based design environment, Windchill is a complete product life cycle management software solution that supports global product development with PDM capabilities. PTC Windchill provides tools to handle data exchange, distributed collaboration, dynamic publishing, and enterprise interoperability. Companies from various industries can utilize PTC Windchill to manage both service and product life cycles.

Key Features of PTC Windchill
- Program management
- Regulatory compliance
- Quality and reliability management
- Change and configuration management
- Requirement capture and management
- System design
- Detailed design
- Verification and validation
- Variant design and generation
- Design outsourcing
- Early sourcing tools
- Component and supplier management
- Manufacturing process management
- Manufacturing outsourcing
- Technical information creation and delivery

Aras innovator
Aras innovator offers an integrated solution for PDM, PLM, and more all on a single platform. The system manages the creation, change, and archive of all information in a centralized or distributed data repository and provides easy access to all users via secure visual collaboration capabilities. Parts and BOM items are automatically extracted and linked to corresponding CAD and documents.

Enovia PDM
Enovia PDM is the collaborative innovation application. It supports bill of material management, configuration and change management, design management, global product development, product planning, program management, and so on.

SmarTeam PDM
SmarTeam was developed by an Israel company Smart Solutions Ltd. It is a PDM workflow software, which later also developed TeamWorks as an integrated workflow solution for SolidWorks. In 1999, the company Smart Solutions Ltd. was acquired by Dassault Systemes.

CAXA-PDM
Caxa-PDM is a part of the collaborative PLM solution, which has been developed by Beijing Caxa company for the product design and management applications.

Caxa-PDM has the functions of engineering document management, product configuration management, CAD integration, workflow, electronic signing, notes in red, project management, BOM, engineering changing management, coding management, and integrating ERP.

6.4 Benefits of PDM systems

People who benefit from the knowledge management and reporting capabilities of PDM systems include project managers, design and manufacturing engineers, sales people, buyers, and quality assurance teams. PDM systems allow companies to:
- Find the correct data quickly
- Get up-to-date bills of materials
- Improve productivity and reduce cycle time
- Reduce development errors and costs
- Meet business and regulatory requirements
- Optimize operational resources
- Facilitate collaboration between global teams
- Provide the visibility needed for better business decision-making
- Improve interdisciplinary collaboration
- Reduce product development cycle time
- Reduce complexity of accessing the information of a company
- Improve project management
- Improve life cycle design
- Enforce supply chain collaboration

With today's integrated systems, PDM/PLM is crucial throughout the organization. It is important to create product data that meet the needs of various groups within the organization. The product data management system allows the company to integrate all aspects of product data with other operations.

A PDM system lets you track the fundamental information required to manufacture components, sub-assemblies, and end item products. PDM is easy to integrate with other systems such as CAD, CAPP, CAM, and ERP. In general, product data management includes bills of material, routing, work centers, and engineering change management.

6.5 New development of PDM

With more applications of PDM technology, its development will get more and more attention. The development of new technology is the driving force of a new generation of PDM. With the appearance and implication of some relative technologies, PDM presents various new development tendencies. Many new products have more major improvement than the previous products. PDM software is evolving with the incorporation of on-demand cloud, big data, mobile, and social elements.

Object-oriented database management

With the development of database technology, the database system has transformed from relation database to object-oriented database. ENOVIApm, formerly known as IBM product manager, is a widely implemented PDM system. ENOVIApm is an integrated suite of object-oriented applications designed to control product and process data in manufacturing organizations of all sizes.

Web-based PDM

Based on web technology, PDM on the internet can be considered as a natural extension of this distributed arrangement. The web browser essentially becomes a new kind of client, one that is uniform across all applications. Internet and web-technology are reshaping the PDM software business. Today, HTML/Java-based web user interfaces and web-based server access are increasingly becoming the norm for PDM systems.

PTC Windchill's browser-based user interface uses standard HTML for bidirectional communication of forms-based information and Java applets to deliver interactive application capabilities. To ease client-side administration, these capabilities are delivered from standard web servers to the web browser using HTTP in a just-in-time, administration-free fashion. Windchill's server technology is based upon Java, and the server process encapsulates the business behavior of product data in an open, flexible architecture, making that data available to various web pages, Java applets, or external systems via HTML, Java RMI, CORBA, or COM technologies.

STEP-based PDM

The STEP AP 242 functions for PDM interoperability. STEP AP 242 PDM data model is based on the convergence of STEP AP 203 and STEP AP 214, and is also shared with STEP AP 209, AP 210, AP 233, and AP 239 (list not exhaustive). Final activities were carried out in 2013 to ensure the PDM harmonization of STEP AP 242 Business Object Model and STEP AP 239 PLCS Platform Specific Model (PSM).

CORBA

The Common Object Request Broker Architecture (CORBA) is a standard defined by the Object Management Group (OMG) designed to facilitate the communication of systems that are deployed on diverse platforms. CORBA enables collaboration between systems on different operating systems, programming languages, and computing hardware. CORBA uses an object-oriented model, although the systems that use CORBA do not have to be object oriented.

From PDM to PLM

Product life cycle management is an information management system that can integrate data, processes, business systems, and, ultimately, people in an extended enterprise. PLM software allows you to manage this information throughout the entire life cycle of a product efficiently and cost-effectively, from concept, design, and manufacture, through service and disposal.

PLM can be viewed as both an information strategy and as an enterprise strategy. As an information strategy, it builds a coherent data structure by consolidating systems. As an enterprise strategy, it lets global organizations work as a single team to design, produce, support, and retire products, while capturing best practices and lessons learned along the way. It empowers your business to make unified, information-driven decisions at every stage in the product life cycle.

Cloud -based PDM

Cloud based PDM is a collaborative engineering data management platform, where all the information is stored in the cloud environment. Cloud-based PDM gets rid of all the hassle of managing the infrastructure required to run a PDM system. Only an individual is assigned to manage the company's online assets and take on the role of PDM administrator: user logins, workflow creation, and training. But, the PDM administrator doesn't have to worry about maintenance packs or updates and no additional IT resources are required.

For some large companies, cloud-based PDM systems allow for projects to be created. Paid users can join as many projects as the terms and conditions of the monthly license allow. One user, one login, multiple projects, each neatly segregated and the project owner controls who has access. The user only interfaces through a single client application, reducing the amount of training needed and reducing confusion. The PDM administrator only has a single environment to maintain as well, so any changes to workflows only need to be applied once and the benefit flowed down to the entire organization.

6.6 Summary

Originally, PDM products organized CAD files and provided version control and check-in/check-out access control. Over time, they took on other tasks, including the management of change orders and bills of material.

The ideal PDM system is accessible by multiple applications and multiple teams across an organization, and supports business-specific needs. Choosing the right PDM software can provide a company in any industry with a solid foundation that can be easily expanded into a full PLM platform. At its core, a PDM system provides solutions for secure data management, process enablers, and configuration management.

This chapter introduces the concept of PDM. The main functions of a PDM system are described. Several popular PDM systems are introduced. The benefits of PDM system are summarized.

Extended reading

For further reading, you are suggested to read the help information for Solidwork Enterprise PDM system at the website on:

http://help.solidworks.com/2013/English/EnterprisePDM/FileExplorer/c_Welcome_to_SolidWorks_Enterprise_PDM.htm.

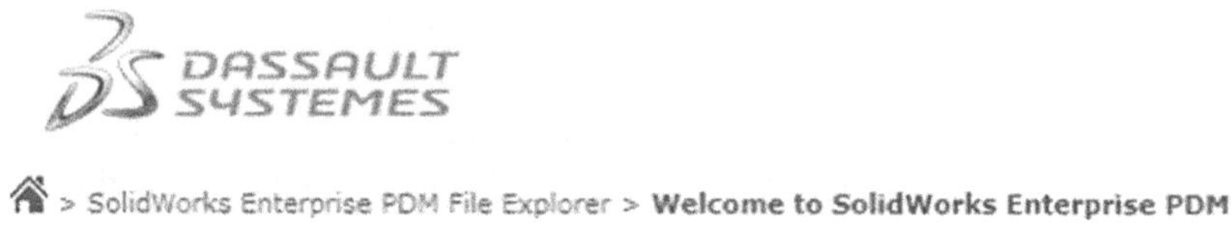

You can also visit the following SAP website to learn something new from SAP Product Data Management.

http://help.sap.com/saphelp_46c/helpdata/en/8a/1a1cf74e4211d182be0000e829fbfe/content.htm

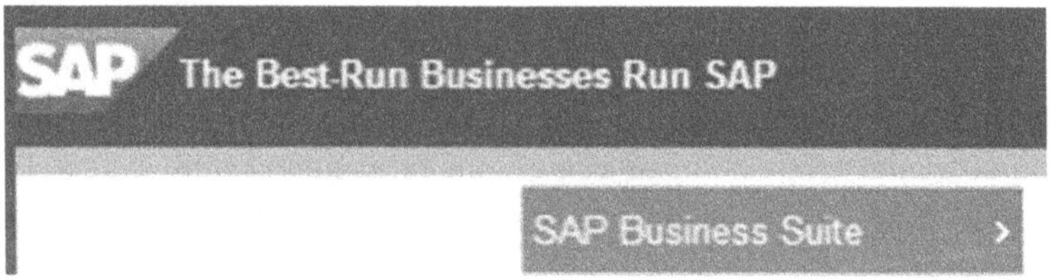

Questions

1. What is product data management?
2. What are the main functions of a PDM system?
3. What is data vault? How many types of data vault?
4. How is a data vault operated?
5. What is workflow management?
6. What is the bill of materials?
7. What is the product configuration?
8. What is the benefit of implementing the PDM system?
9. How does a PDM system help the integration of CAD/CAPP/CAM?

Tasks

Visit some local companies that have implemented the PDM systems to manage the engineering documents. Based on the knowledge learned in this chapter, write a report.

References

Xu X., Integrating advanced computer aided design, manufacturing, and numerical control, principles and implementations. Information Science Reference, 2007.

http://blog.csdn.net/biili/article/details/710504

http://www.ap242.org/pdm-interoperability

http://www.cadalyst.com/data-management/new-expectations-drive-development-pdm-collaboration-features-23107

http://www.caxa.com/plm/pdm/pdm3.html

http://www.3ds.com/products-services/enovia/

7 Concurrent engineering and collaborative design

Questions before you read

1. Do you know the procedure of product design?
2. What are the general steps for sequential product design?
3. What are the disadvantages of a sequential product design?
4. Do you have some ideas about concurrent design engineering?
5. Do you know about the development in concurrent engineering technology?
6. What is collaborative design?

The goal of this chapter is to introduce the concept of concurrent engineering. The continuous improvement is been done with various quality control approaches. The design and manufacturing of Boeing 777 are taken as an example in this chapter.

7.1 Introduction

Concurrent engineering (CE) as a concept has been known for many years. In many industries, it is widely understood as a common term for developing new products in a cross functionally integrated manner primarily by using multifunctional teams and overlapping of sequential activities. The concept of concurrent design, taking into account the manufacturability at the design stage, has further strengthened the need for improvement in the present techniques of part description. The concept demands that manufacturing aspects should be integrated with the design process, and an acceptable design should not only fulfill the functional requirements but also be economical and easy to produce in terms of cost and manufacturing time.

Definition of concurrent engineering

CE is the earliest possible simultaneous work of experts from various functions in an enterprise, concerned with producing a specific product, in order to achieve high quality, functionality and manufacturability in the shortest time, for a minimum cost. CE is primarily an expression for the desire to increase the competitiveness by decreasing the product lead time, while improving its quality and cost. According to the business dictionary.com, CE is defined as follows:

Integrated approach to product design that takes into account all stages of a product's life cycle from design to disposal – including costs, quality, testing, user needs, customer support, and logistics.

–Courtesy to Business Dictionary.com

The main premise of CE method is the integration of product design, process planning, and manufacturing processes, or the integration of computer-aided design (CAD)/computer-aided process planning (CAPP)/computer-aided manufacturing

https://doi.org/10.1515/9783110573091-007

(CAM). CE relies on well-founded methods, efficient tools, and a dedicated implementation team for its success. The work way of CE requires various engineering activities in the product design, process planning, and manufacturing processes to be integrated and performed in a parallel manner rather than in a sequential manner.

CE is a method by which several teams within an organization work simultaneously to develop new products and services. By concurrently engaging in multiple aspects of development, the amount of time involved in getting a new product to the market has significantly decreased. In markets where customers value time compression, fast-cycle developers have a distinct advantage. Additionally, in many high-technology areas, product-technology performance is continuously increasing and price levels are dropping almost daily. In such areas, a company's ability to sustain its competitive edge largely depends on the timely introduction of new or improved products and technologies. CE is a key method for meeting this need of shortening a new product's time-to-market.

Why CE?

The recent advances in telecommunication and transportation, global alliances among enterprises, and changing customer needs characterize a rapidly emerging global market economy. Products entering this market are designed and manufactured across geographical boundaries and distributed and marketed worldwide. In addition to a worldwide competition, companies are also faced with shrinking time-to-market for new products. This is the elapsed time from product concept to its actual availability on store shelves. During this period, the product goes through several stages that collectively define the product life cycle.

The most crucial stage in the product life cycle is the design stage. Any mistake in the design stage can cost a lot in terms of engineering changes and its impact on manufacturing, delays in product release to the market with potential loss of the market, and product recalls in the case of a released product with significant financial losses. Hence, there should be special emphasis on the product design to ensure that the product can flawlessly reach the market and in the fastest time possible. Getting it right the first time, which is all the more vital in a global market, can be implemented only with a good design.

A company's competitive advantage in the global market depends on its ability to produce high-quality products at a low price in the shortest time. It should also be able to continuously innovate to both the product and the processes that produce the product in order to respond quickly to the market changes and reduce the risk of failure. The CE is a superior tool in achieving these objectives.

Comparison between sequential approach and concurrent engineering

CE is not a new concept. It was used in Japan in the early 1960s and was promoted and popularized by the United States in the 1990s. It is slightly different from the traditional design approach. A comparison between concurrent engineering and the traditional sequential approach produces some interesting conclusions.

The traditional sequential approach

Traditionally, products were designed and manufactured following the sequential engineering methods, where people from different departments work one after the other on successive phases of development. This method of production is in a linear format. The different steps are done one after another, with all attention and resources focused on that one task. After it is completed, it is left alone and everything is concentrated on the next task. The product is first completely defined by the design department, then the process planning department takes over, and finally the manufacturing department defines the manufacturing process. This is a long time process, and often leads to a lot of design changes as the prototype testing begins, due to production problems, delays, or design flaws. Therefore, this is a slow and expensive approach, often leading to low-quality and less competitive products. The traditional sequential approach is still used by many companies. As illustrated in Figure 7.1, the information flow in the sequence is linear, with little interaction between the groups.

The CE approach

The concurrent engineering approach is illustrated in Figure 7.2, which is a systematic approach to the integrated, concurrent design of products and their related processes, including manufacture and support. This approach is intended to cause the

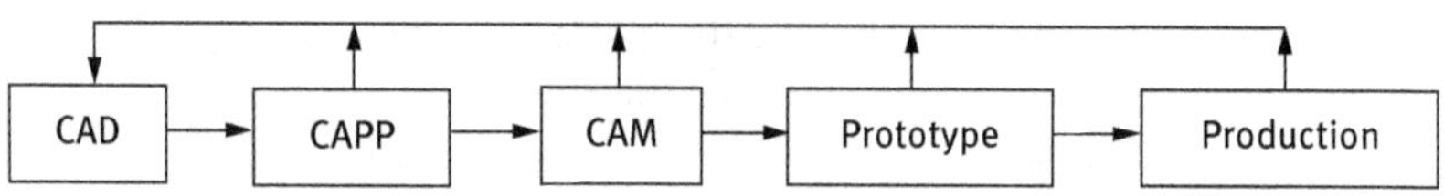

Figure 7.1: The traditional sequential engineering.

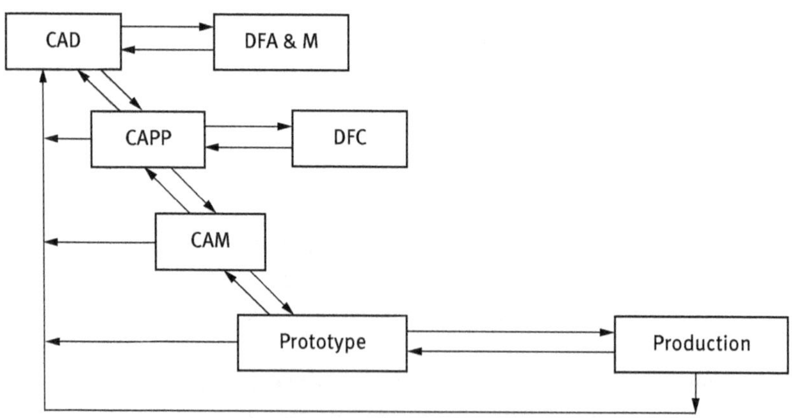

Figure 7.2: The modern concurrent engineering.

developers, from the outset, to consider all elements of the product life cycle from conception through disposal, including quality, cost, schedule, and user requirements. This results in the product development team clearly understanding what the product requires in terms of mission performance, environmental conditions during operation, budget, and scheduling.

In this method, several teams within an organization work simultaneously to develop new products and services, and this therefore allows a more streamlined approach. Decision-making involves full-team participation and involvement. The team often consists of product designers, manufacturing engineers, marketing personnel, purchasing, finance, and suppliers, and the role of the leader is to supply the basic foundation and support for change, rather than to tell the other team members what to do. In CE, different tasks are tackled at the same time, and not necessarily in the usual order. This means that information found out later in the process can be added to earlier parts, improving them, and also saving a lot of time. Examples from companies using CE techniques show significant increases in overall quality, 30–40% reduction in project times and costs, and 60–80% reductions in design changes after release.

Model of CE

A typical model of CE in the realization of a product is shown in Figure 7.3. The CE model relies on a CE team that is responsible for the total product life cycle, from a conceptual idea to the finished product. Such a team brings together design, engineering, and manufacturing expertise.

The market and cost analysis phase defines the new conceptual idea in terms of market requirements and suggests selling price and production cost targets, which then serve as input to the design wheel. The domain experts from different departments form the CE team; they engage in the design phase to ensure that the functional goals are met while adhering to cost targets and other constraints.

The marketing department typically initiates new product conceptual ideas. Such ideas are based on market research, where the customers' needs are studied. The marketing department is not totally independent in conducting this marketing research. Various other departments contribute information such as the manufacturability of a certain product proposal and help the marketing department in formulating a more precise idea for a new product. Eventually, the idea that is presented to the design team has more clearly defined design goals and objectives. An output of the marketing analysis is the setting of the selling price and production cost targets, as shown in the left block of Figure 7.3.

The principal difference in the concurrent design engineering versus the traditional design engineering is in the design approach. At the CE design stage, a suggested design is submitted to the CE team. The members of CE team involved in the design stage are the domain experts from the marketing, designing, engineering, manufacturing, sales, packaging, inspection, service, assembly, and environmental

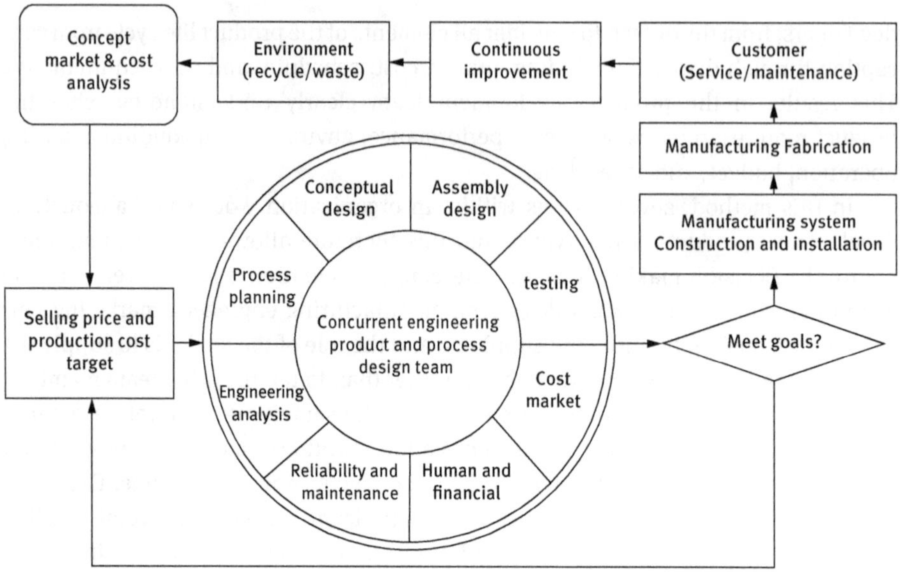

Figure 7.3: A model of CE.

departments. They will improve the design of the product with implementation of their knowledge by working simultaneously with the design team.

A good feature of this CE model is that the CE team not only designs the product but also designs the process for the manufacturing of the product. The team decides what kind of equipment is going to be used, the layout of the machines, etc. In the traditional model, at this stage, only people from the design department are involved, the big decisions are mostly depended on the designer, and the manufacturing of the product is left to the manufacturing engineers.

A comparison of the CE model and the traditional model of product realization is shown in Figure 7.4. As can be seen, there is a huge time saving when CE is implemented in the design-to-manufacturing cycle of the product realization. In addition, the CE method does not lead into problems of implementing the design in manufacturing such as costly engineering changes. This will result in reducing the overall product cost.

In the traditional model, once the design is made, the manufacturing departments that are involved in the product realization are expected to follow it, although they have very little input in the design of the product. A frequently asked question is, *how good the design can be without the involvement of domain experts?* Very often, the design team in the traditional model has little knowledge and skills to make a product that will be functional, of high quality, and easy to manufacture.

After the design team completes its task, the production processes are designed based on the product design. Therefore, if the product is poorly designed, the ensuing

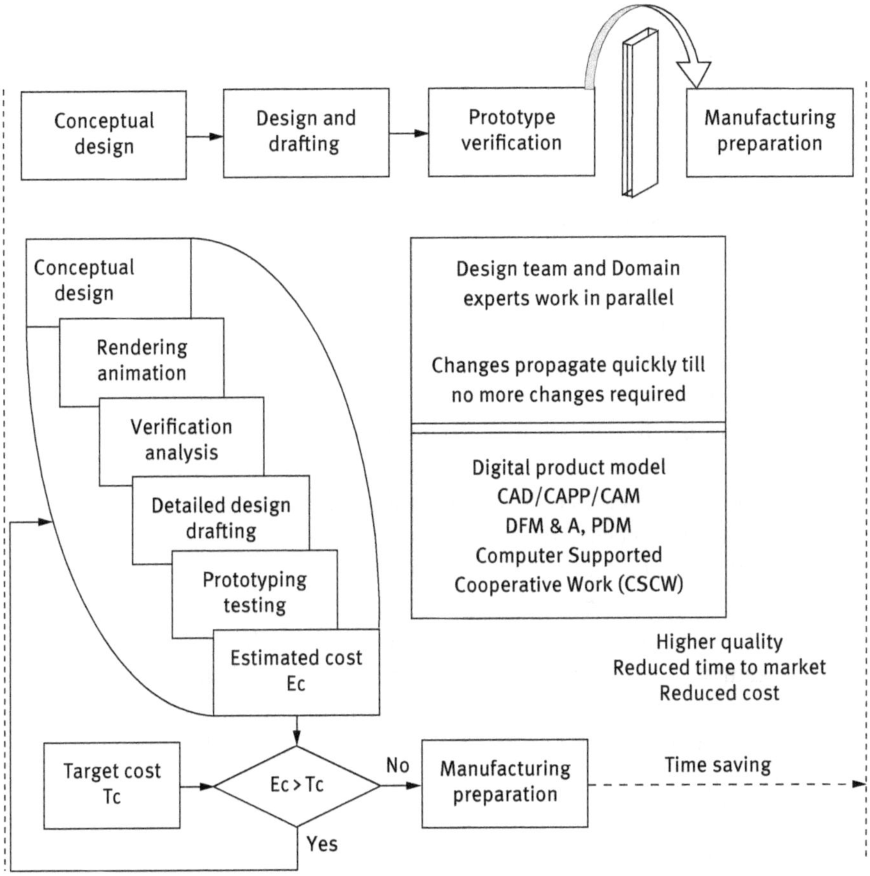

Figure 7.4: A comparison between traditional design and CE.

processes will be poorly designed, too. For example, if the manufacturing department has a part that is difficult to manufacture due to the poor design, a considerable time will be expended in order to manufacture the part. To accomplish this, sometimes, the manufacturing department introduces changes to the original design such as either updating the part tolerances or changing the number of parts in the design. At the same time, the changes in the product design may not be either communicated to others in the product realization process or too late to prevent decisions that are based on the original product design. At any rate, the traditional model is vulnerable to a costly and error-prone product realization.

Figure 7.4 shows that while the CE design method begins with a cost target for the product, the traditional method has no such benchmark and does not consider the cost at first. Following the design stage, the CE team compares the derived cost of the product design to the targeted cost. Only if the estimated cost is lower or equal to the targeted cost, the manufacturing of the product can begin. Such design discipline is essential to ensure that the price of the product is competitive in the market.

It is obvious that by following the CE model all the disadvantages of the traditional model can be avoided. The difference in the two approaches will be more evident upon study of the overall production cost in the product's life cycle.

Benefits and advantages of CE

CE provides many benefits over traditional sequential engineering, including lower manufacturing and production costs, improved quality of resulting end products, and increased accuracy in predicting and meeting project plans, schedules, timelines, and budgets. Because the multidisciplinary teams working together early in the process can make informed decisions about cost, quality, process and product issues, trade-offs can be made between design features, part manufacturability, assembly requirements, material needs, reliability issues, serviceability requirements, and cost and time constraints. Any differences are usually reconciled early in the design process, leading to increased efficiency and performance, higher reliability in the product development process, reduced defect rate, and ultimately a faster time to market that results in increased market share. This also means faster reaction times in responding to the rapidly changing market, which in turn fosters increased customer satisfaction and a higher return on investments due to the reduced labor and resource requirements, improved inventory control, and scheduling. Improved communication between individuals and departments within the company also encourages cohesiveness and a more pleasant working environment, which in turn can positively affect productivity of the workforce.

CE is not a trivial process to apply; therefore, companies must be careful in using this approach. To be successful, they should initially compare themselves with their competitors to set a benchmark, and identify potential performance improvements and realistic targets by analyzing the market and knowing the customers. It is paramount to have the top management's support and to develop a clear strategy and implementation plan, which must be continually reviewed and revised with progress. Individualism should be suppressed within the team, and project leaders must have a clear overall visualization of the project and goals. Cross-functional integration and collaboration need to be established and encouraged, so as to foster team ethics and freely transfer technology and information between individuals and departments. Problems can also arise if the employees have not had any training in teamwork. There needs to be a change in relationships with vendors for the CE methods to work at their best, alongside a focus on process improvement rather than computerization. CE is an evolving process that requires continuous improvement and refinement. This continuous improvement cycle needs business process re-engineering (BPR) in the CE development process.

7.2 Business process re-engineering of product development

Business process re-engineering

BPR is the practice of redesigning the way work is done to better support an organization's mission and reduce costs. Re-engineering focuses on the organization's business processes – the steps and procedures that govern how resources are used to create products and services that meet the needs of particular customers or markets. As a structured ordering of work steps across time and place, a business process can be decomposed into specific activities, measured, modeled, and improved. It can also be completely redesigned or eliminated altogether. Re-engineering identifies, analyzes, and redesigns an organization's core business processes with the aim of achieving significant improvements in critical performance measures, such as cost, quality, service, and speed.

Shewhart cycle (PDCA)

PDCA (plan–do–check–act or plan–do–check–adjust) is an iterative four-step management method used in business re-engineering for the control and continuous improvement of processes and products. PDCA was made popular by Dr W. Edwards Deming, who is considered to be the father of modern quality control. However, PDCA is always referred as the "Shewhart cycle," as shown in Figure 7.5.

The concept of PDCA is based on the scientific method. Shewhart described manufacture under statistical control as a three-step process of specification, production, and inspection. He also specifically related this to the scientific method of hypothesis, experiment, and evaluation. Clearly, Shewhart intended the analyst to take action based on the conclusions of the evaluation. So finally he invented the methods of plan, do, check, and act.

A fundamental principle of the scientific method and PDCA is iteration – once a hypothesis is confirmed (or negated), executing the cycle again will extend the knowledge further. Repeating the PDCA cycle can bring us closer to the goal, usually a perfect operation and output. Deming continually emphasized iterating toward an improved system; hence, PDCA should be repeatedly implemented in spirals of increasing knowledge of the system that converge on the ultimate goal, each cycle

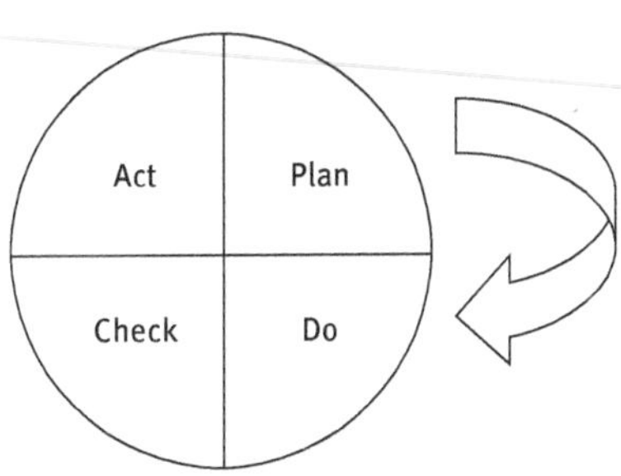

Figure 7.5: The Shewhart cycle.

closer than the previous. One can envision an open coil spring, with each loop being one cycle of the scientific method – PDCA, and each complete cycle indicating an increase in our knowledge of the system. This approach is based on the belief that our knowledge and skills are limited, but improving. Especially at the start of a project, key information may not be known; the PDCA provides feedback to justify our guesses and increase our knowledge. With the improved knowledge, we may choose to refine or alter the goal. Certainly, the PDCA approach can bring us closer to whatever goal we choose.

Rate of change, that is, rate of improvement, is a key competitive factor in today's world. PDCA allows for major breakthroughs in performance, as well as frequent small improvements. The scientific method of PDCA applies to all sorts of projects and improvement activities.

Plan

Establish the objectives and processes necessary to deliver results in accordance with the expected target or goals. By establishing result expectations, the completeness and accuracy of the spec is also a part of the targeted improvement.

Do

Implement the plan, execute the process, and make the product. Collect data for charting and analysis in the following steps.

Check

Study the actual results (measured and collected in "Do" step) and compare against the expected results (targets or goals from the "Plan") to ascertain any differences. Look for deviation in implementation from the plan and also look for the appropriateness and completeness of the plan to enable the execution, that is, "Do." Charting data can make this much easier to see trends over several PDCA cycles and to convert the collected data into information. Information is what you need for the next step.

Act

If the CHECK shows that the PLAN that is implemented in DO is an improvement to the prior standard, then that becomes the new standard for how the organization should ACT going forward. If the CHECK shows that the PLAN that is implemented in DO is not an improvement, then the existing standard will remain in place. In either case, if the CHECK showed something different than expected (whether better or worse), then there is some more learning to be done and that will suggest potential future PDCA cycles.

Note, sometime the "A" also refer to as "Adjust." This helps to understand that the fourth step is more about adjusting/correcting the difference between the current state and the planned state instead of thinking that the "A" is all about action and implementation. For those who prefer "Adjust" be aware that the thing being adjusted are the standards. Those adjusted standards will now inform and guide the next "Plan" phase. Sometimes "Adjust" may lead some to believe that it is the "Plan" itself that is being adjusted. The new plan may well look a bit different, but it is now because it is formulated while resting on a new set of standards. This is subtle, but important.

Six Sigma

Six Sigma is a disciplined, data-driven approach and methodology for eliminating defects in any process – from manufacturing to transaction and from product to service. In Six Sigma programs, the PDCA cycle is called "define, measure, analyze, improve, control" (DMAIC). Sometime, it is also called "define, measure, analyze, design, verify" (DMADV).

The statistical representation of Six Sigma quantitatively describes how a process is performing. To achieve Six Sigma, a process must not produce more than 3.4 defects per million opportunities. A Six Sigma defect is defined as anything outside of customer specifications.

The fundamental objective of the Six Sigma methodology is the implementation of a measurement-based strategy that focuses on process improvement and variation reduction through the application of Six Sigma improvement projects. This is accomplished through the use of two Six Sigma submethodologies: DMAIC and DMADV. The Six Sigma DMAIC process is an improvement system for existing processes falling below specification and looking for incremental improvement. The Six Sigma DMADV process is an improvement system used to develop new processes or products at Six Sigma quality levels. It can also be employed if the current process requires more than just incremental improvement. Both Six Sigma processes are executed by Six Sigma Green Belts and Six Sigma Black Belts, and are overseen by Six Sigma Master Black Belts.

General Electric, one of the most successful companies implementing Six Sigma, has estimated benefits on the order of $10 billion during the first five years of implementation. General Electric first began Six Sigma in 1995. Since then, thousands of companies around the world have discovered the far reaching benefits of Six Sigma.

Many frameworks exist for implementing the Six Sigma methodology. Six Sigma consultants all over the world have developed proprietary methodologies for implementing Six Sigma quality, based on the similar change management philosophies and applications of tools.

Process planning and management for product development

Gantt chart

A Gantt chart is a type of bar chart that illustrates a project schedule. Gantt charts illustrate the start and finish dates of the terminal elements and summary elements of a project. Terminal elements and summary elements comprise the work breakdown structure of the project. Modern Gantt charts also show the dependency relationships between activities, such as precedence network.

Table 7.1: The tasks for a Gantt chart.

Activity	Predecessor	Time estimates			Expected time (TE)
		Opt. (O)	Normal (M)	Press (P)	
A	Null	2	4	6	4.00
B	Null	3	5	9	5.33
C	A	4	5	7	5.17
D	A	4	6	10	6.33
E	B, C	4	5	7	5.17
F	D	3	4	8	4.5
G	E	3	5	8	5.17

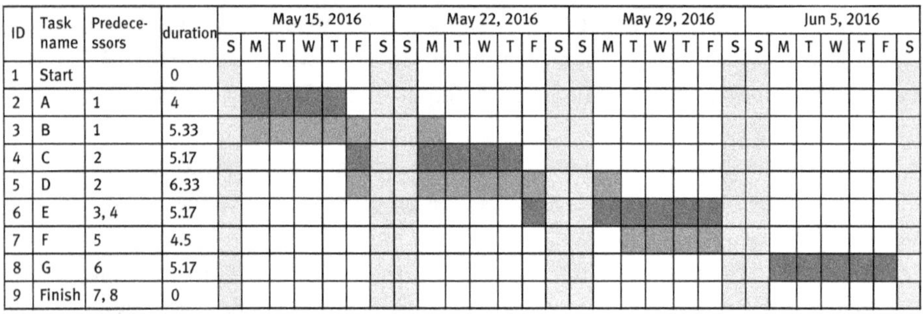

Figure 7.6: An example of a Gantt chart.

In Table 7.1, there are seven tasks, labeled A through G. Some tasks can be done concurrently (A and B) while others cannot be done until their predecessor task is complete (C cannot begin until A is complete). Additionally, each task has three time estimates: the optimistic time estimate (O), the most likely or normal time estimate (M), and the pessimistic time estimate (P). The expected time (TE) is estimated using the β probability distribution for the time estimates, using the formula $(O + 4M + P) \div 6$.

Once this step is complete, one can draw a Gantt chart or a network diagram.

Gantt charts can be used for scheduling generic resources as well as for project management. They can also be used for scheduling production processes. Gantt charts are one of the important tools used in CE. Figure 7.6 is an example of a Gantt chart.

Critical path method

The critical path method (CPM) is a project modeling technique developed in the late 1950s. CPM is commonly used with all forms of projects, including construction, aerospace and defense, software development, research projects, product development, engineering, and plant maintenance, among others. Any project with interdependent activities can apply this method of mathematical analysis.

The essential technique for using CPM is to construct a model of the project that includes the following:

1. a list of all activities required to complete the project (typically categorized within a work breakdown structure);
2. the time (duration) that each activity will take to complete;
3. the dependencies between the activities; and
4. logical end points such as milestones or deliverable items.

Using these values, CPM calculates the longest path of planned activities to logical end points or to the end of the project, and the earliest and latest that each activity can start and finish without making the project longer. This process determines which activities are "critical" (i.e., the longest path) and which have "total float" (i.e., can be delayed without making the project longer). In project management, a critical path is the sequence of project network activities that add up to the longest overall duration, regardless if that longest duration has float or not. This determines the shortest time possible to complete the project. There can be "total float" (unused time) within the critical path.

For example, if a project is testing a solar panel and task "B" requires "sunrise," there could be a scheduling constraint on the testing activity so that it would not start until the scheduled time for sunrise. This might insert dead time (total float) into the schedule on the activities on that path prior to the sunrise due to needing to wait for this event. This path with the constraint-generated total float would actually make the path longer, with total float being part of the shortest possible duration for the overall project. In other words, individual tasks on the critical path prior to the constraint might be able to be delayed without elongating the critical path; this is the "total float" of that task. However, the time added to the project duration by the constraint is actually critical path drag, which is the amount by which the project's duration is extended by each critical path activity and constraint.

CPM analysis tools allow a user to select a logical end point in a project and quickly identify its longest series of dependent activities (its longest path). These tools can display the critical path as a cascading waterfall that flows from the project's start to the selected logical end point.

Activity-on-node diagram (program evaluation and review technique)

An activity-on-node diagram shows critical path schedule, along with total float and critical path drag computations. Although the activity-on-arrow diagram (program evaluation and review technique, "PERT chart") is still used in a few places, it has generally been superseded by the activity-on-node diagram, where each activity is shown as a box or node and the arrows represent the logical relationships going from predecessor to successor as shown in the "activity-on-node diagram" (Figure 7.7).

In this diagram, activities A, B, C, D, and E comprise the critical or longest path, while activities F, G, and H are off the critical path with floats of 15, 5, and 20 days,

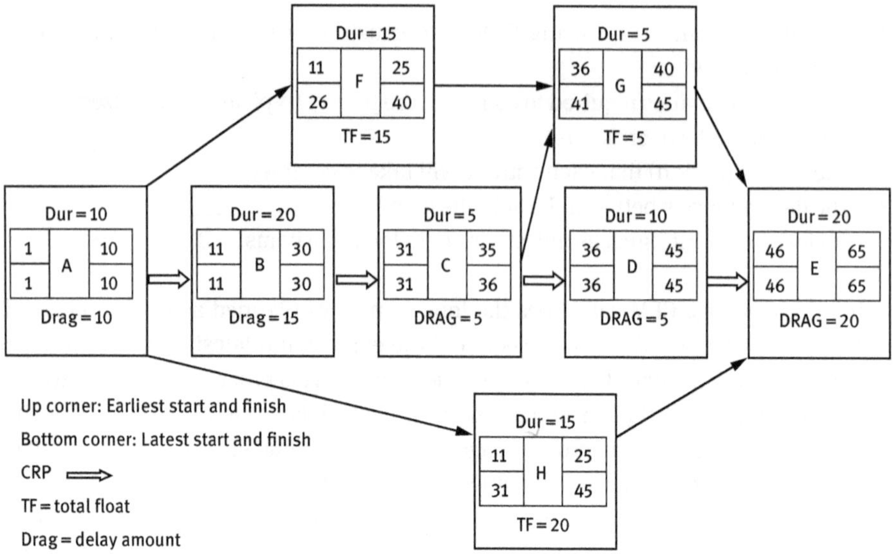

Figure 7.7: Activities-on-node diagram and PERT.

respectively. Although activities that are off the critical path have float and are therefore not delaying completion of the project, those on the critical path will usually have critical path drag, that is, they delay project completion. The drag of a critical path activity can be computed using the following formula:

1. If a critical path activity has nothing in parallel, its drag is equal to its duration. Thus, A and E have drags of 10 and 20 days, respectively.
2. If a critical path activity has another activity in parallel, its drag is equal to whichever is less: its duration or the total float of the parallel activity with the least total float. Thus, since B and C are both parallel to F (float of 15) and H (float of 20), B has a duration of 20 days and drag of 15 days (equal to F's float), while C has a duration of only 5 days and thus drag of only 5. Activity D, whose duration is 10 days, is parallel to G (float of 5) and H (float of 20) and therefore its drag is equal to 5, the float of G.

These results, including the drag computations, allow managers to prioritize activities for the effective management of project completion, and to shorten the planned critical path of a project by pruning critical path activities, by "fast tracking" (i.e., performing more activities in parallel), and/or by "crashing the critical path" (i.e., shortening the durations of critical path activities by adding resources).

Critical path drag analysis has also been used to optimize schedules in processes outside of strict project-oriented contexts, such as to increase manufacturing throughput by using the technique and metrics to identify and alleviate delaying factors and thus reduce assembly lead time.

7.3 Key technologies of CE

There are some key technologies of CE, such as digital product modeling and product data management (PDM), design for manufacturing and assembly (DFA&M), integration of CAD/CAPP/CAM, and collaborative works.

Digital product modeling and PDM

The designer generally has access to models that others are concurrently working on. For example, several people may be designing one machine that has many parts. New parts are added to an assembly model as they are created. The design evolution is visible to everyone involved. Depending on the system, it might be necessary for the users to acquire the latest versions saved for each individual component to update the assembly.

So the geometry model, feature model, and assembly model should be digital. They are integrated modes. These models should be managed with a PDM system. The PDM system needs to be integrated with enterprise resource planning (ERP). Standard for the exchange of product (STEP)-based information integration systems are one of the enable technology of CE.

In CE, design technologies are utilized that foster efficient cross-disciplinary analysis, experimentation, and representation of new product designs. Some examples of these technologies include the following: 3D CAD systems, rapid prototyping techniques, rapid tooling, and rapid testing techniques, as well as techniques that enable the representation of product designs in a virtual context. These design technologies are important because of the key information they convey: their 3D character allows the expert to interpret design features in a more effective and efficient way.

All of these technologies contribute to the reduction of interpretation asymmetries between the experts involved, as well as to fast-cycle design and development, because they allow for high-speed iterations of analysis and experimentation on both concepts and models of the product. Thus, they modify traditional project management approaches by allowing for more systematic and flexible experimentation and iteration to be included throughout the project's design and development process. In fact, the time and cost incurred by the development and construction of prototypes generally are reduced by factors of 2–5 when using digital (e.g., 3D CAD) and physical (e.g., rapid prototyping) technologies. These tools have become important enabling factors in the CE environment. Without their implementation and further upgrading, CE might never be able to realize its full potential in terms of design cost and lead-time optimization.

Design for manufacturing and assembly

DFA&M is a technique for reducing the cost of a product by breaking the product down into its simplest components. All members of the design team can understand the product's assembly sequence and material flow early in the design process.

The road map or the methodology of DFM&A is described as follows:
1. Form a multifunctional team.
2. Establish the product goals through competitive bench-marking.
3. Perform a design for assembly analysis.
4. Segment the product into manageable subassembly or levels of assembly.
5. As a team, apply the design for assembly principles.
6. Use creativity techniques to enhance the emerging design.
7. As a team, evaluate and select the best ideas.
8. Ensure economical production of every piece of part.
9. Establish a target cost for every part in the new design.
10. Start the detailed design of the emerging product.
11. Apply design for producibility guidelines.
12. Reapply the process at the next logical point in the design.
13. Provide the team with a time for reflection and sharing results.

Usually a DFM&A checklist is used for the development of a new product, as listed in Table 7.2.

Design for assembly (DFA) is a process by which products are designed with ease of assembly in mind. If a product contains fewer parts, it will take less time to assemble, thereby reducing assembly costs. In addition, if the parts are provided with features that make it easier to grasp, move, orient, and insert them, this will also reduce assembly time and assembly costs. The reduction of the number of parts in an assembly has the added benefit of generally reducing the total cost of parts in the assembly. So by analyzing the product structure and assemble and tolerance allocation, the assembly will get an optimized solution. An assemble plan design is often optimized by simulation of assemble.

Design for manufacturing (DFM) is the general engineering art of designing products in such a way that they are easy to manufacture. DFM describes the process of designing a product in order to facilitate the manufacturing process to reduce manufacturing costs. DFM will allow potential problems to be fixed in the design phase, which is the least expensive place to address them. Other factors may affect the manufacturability such as the type of raw material, the form of the raw material, dimensional tolerances, and secondary processing such as finishing. After using DFM, the part structure is reasonable and feasible to be manufactured.

Some time, design for cost (DFC) will use cost-estimating model to reduce the cost of the product. DFA&M and DFC can help reduce cost, improve the product quality, and reduce the developing time.

Collaborative work in CE

With the use of computer in engineering, CAD/CAPP/CAM systems are integrated by using STEP and STEP-NC. It is easy to exchange product data among different systems. PDM system may archive all the information of developing a new product. Hence, a collaborative work environment can be established based on the information technology,

Table 7.2: DFM&A new products checklist.

Design for manufacturing and assembly consideration	**Yes**	**No**
Design for assembly analysis completed:	☐	☐
Has this design been analyzed for minimal part count?	☐	☐
Have all adjustments been eliminated?	☐	☐
Are more than 85% common parts and assemblies used in this design?	☐	☐
Has assembly sequence been provided?	☐	☐
Have assembly and part reorientations been minimized?	☐	☐
Have more than 96% preferred screws been used in this design?	☐	☐
Have all parts been analyzed for ease of insertion during assembly?	☐	☐
Has all assembly interference been eliminated?	☐	☐
Have location features been provided?	☐	☐
Have all parts been analyzed for ease of handling?	☐	☐
Have part weight problems been identified?	☐	☐
Have special packaging requirements been addressed for problem parts?		
Are special tools needed for any assembly steps?	☐	☐
Ergonomics considerations	**Yes**	**No**
Does design capitalize on self-alignment features of mating parts?	☐	☐
Have limited physical and visual access conditions been avoided?	☐	☐
Does design allow for access of hands and tools to perform necessary assembly steps?	☐	☐
Has adequate access been provided for all threaded fasteners and drive tooling?	☐	☐
Have all operator hazards been eliminated (sharp edges)?	☐	☐
Design for manufacturing and considerations	**Yes**	**No**
Have all unique design parts been analyzed for producibility?	☐	☐
Have all unique design parts been analyzed for cost?	☐	☐
Have all unique design parts been analyzed for their impact of tooling or mold cost?	☐	☐
Design for assembly process and considerations	**Yes**	**No**
Has assembly tryout been performed prior to scheduled prototype build?	☐	☐
Have assembly views and pictorial been provided to support assembly documentation	☐	☐
Has opportunity defects analysis been performed on process build?	☐	☐
Has products cosmetics been considered (paint match, scratches)?	☐	☐
Wire management	**Yes**	**No**
Has adequate panel pass through been provided to allow for easy harness/cable routing?	☐	☐
Have harness/cable supports been provided?		
Have keyed connector been provided at all electrical interconnections?	☐	☐
Are all harness/cables long enough for ease of routing, tie down, plug in, and to	☐	☐
eliminate strain relief on interconnects?	☐	☐
Does design allow for access of hand and tools to perform necessary wiring operations?	☐	☐
Does position of cable/harness impede air flow?	☐	☐

communication technology, and digital modeling. With the work flow management and constrains management, the designers or engineers from different departments can have some negotiation in CE. Even the history information can be accessed. So the computer-supported cooperative work is also necessary for the CE. Thus, the collaborative work can be carried out in CE.

7.4 Example of Boeing 777-X

Introduction

One of the most famous approaches to CE in the aerospace industry is the Boeing 777-X program. CE acts as a systematic communication between team members to enable consideration of all important product and process information in a timely manner. The Boeing 777-X approach, by means of advanced computer technologies and multifunction teams, brings the initiative of the designer, manufactures, suppliers, and customers into full play. The Boeing 777-X approach seems to be adaptable and has been proven that it is a very effective way to shorten the development time and get better product even if it is a very complex product. When choosing this approach, CE should be applied step by step.

Traditional sequential new product development

In the past, Boeing could design, develop, and manufacture high-quality products that satisfied real needs at competitive prices. This practically guaranteed their commercial success. However, beginning in the early 1990s, this traditional formula radically changed as time-to-market became a vital component of commercial success. In other words, time has become a key driver of competitive success, from design and development to the actual launch of a new product or service.

Traditional project planning and execution has been marked by the definition of objectives and milestones. These goals are met through a progression of networked activities, some of which must be performed sequentially, and others of which may be conducted in parallel. Planning techniques such as PERT, graphical evaluation and review technique, and CPM have been used to support this sequencing of tasks and activities. However, until the beginning of the 1990s, time compression was not a major issue in the new product development environment. In the planning and scheduling of tasks and activities, any time compression concerns are only implicitly present.

Concurrent new product development

Because time has become a competitive weapon, time pressures have become central to the project-based new product development organization. These pressures have led to the explicit understanding that time compression is a driver of project performance. As a consequence, methods, techniques, and organizational approaches have been designed and developed that allow for time compression needs to be handled in a proper manner. All time-centered approaches have one principle in common: they attempt to maximize the number of major design or development tasks that are performed concurrently, thus the concept of CE.

In a CE environment, even if certain tasks cannot be completely executed at the same time, designers and developers are encouraged to achieve maximum overlap between otherwise sequential activities. In other words, CE aims at achieving throughput time reductions by planning and executing design and development activities in parallel, or by striving for maximum overlap between activities that cannot be completely executed in parallel. For example, when one of the tasks or activities requires information to be partially generated during a previous task or activity.

Therefore, CE is based on the premise that the parallel execution of major design components will decrease the throughput time of projects, thus reducing the time-to-market for new products and services. For example, applying concepts of parallelism during the Boeing 777-X transport design resulted in a time compression of 1.5 years as compared to its predecessor, the Boeing 767. CE allowed the Boeing company to introduce a new airplane in time to limit the advantage of its competitor, Airbus Industrie.

Implementing CE

In a CE environment, teams of experts from different disciplines are formally encouraged to work together to ensure that design progresses smoothly and that all participants share the same current information. The project methods, problem-solving methods and the technologies utilized make up the essential elements through which parallelism in new product design and development can be achieved. Following is a discussion on how each of these elements contributes to CE implementation.

Project methods

Project methods based on teamwork, milestone management, and target-oriented work definition and follow-up are paramount. These methods must also be supported by appropriate senior management commitment and incentive systems. Each team is granted a large degree of autonomy to solve design problems where and when they occur, without much hierarchical intervention. However, management must ensure that the transfer of information between different activities or tasks is smooth and transparent. In addition, the means of experimentation must allow the experts involved to rule out differences in interpretation on the functional and technical design parameters. In other words, for CE to be successful, information and interpretation asymmetries between the experts involved must be avoided whenever possible.

Problem-solving methods

During design and development projects, methods are utilized that foster and support smooth interdisciplinary problem definition and problem solving. Methodologies such as brainstorming open the boundaries of the team to allow for wider ranges of alternative design definitions and solutions to be considered. The use of methodologies like quality function deployment further aids experts from different disciplinary backgrounds to jointly define a product's functional and technical requirements. Activity flow chart methods such as Integrated computer-aided manufacturing DEFinition (IDEF3) allow for detailed planning and monitoring of the different parallel and overlapping activities involved in project execution. Failure mode and effects analysis allows for a systematic investigation of the occurrence and impact of possible flaws in the new product design. The use of design of experiments enables the systematic identification of critical product/process parameters that influence performance. These are just a few of the many supportive methods that can be used in a CE environment.

FlyThru for digital preassembly

Digital preassembly is a kind of processes created on the Boeing 777-X for checking designs. FlyThru software was used as a spinoff of a Boeing advanced computing research project. Engineers were able to view up to 1,500 models in 3D CAD by traversing product data at high speed. FlyThru was rapidly deployed in 1991 to meet the needs of the 777-X for large-scale product visualization and verification. The digital preassembly process has gotten very good results. The Boeing 777-X has had far fewer assembly and systems problems compared to previous airplane programs. Today, FlyThru is installed on hundreds of workstations on almost every airplane program, and is being used on other projects. Their applications have gone far beyond just design review. In many ways, FlyThru is a data warehouse supported by advanced tools for analysis. It is today being integrated with knowledge-based engineering geometry generation tools.

7.5 Summary

In today's business world, a quick effective response to changing market needs is very important if a corporation is to be successful. They must be able to reduce their time to market with an adaptable attitude and decision must be made quickly and correctly the first time around. If the firms waste time repeating tasks, as might happen using sequential methods, they will become less competitive; therefore, CE has emerged as a way of bringing rapid solutions to product design and development process.

This chapter has provided an overview of the why, what, and how to involve in implementing a CE philosophy for the development of new products, services, and processes. It has outlined how introducing overlap during the execution of innovation project tasks and activities has become vital because of competitive pressures that force new product developers to be more time conscious.

Although CE is an important method for handling the time pressures that occur during new product development, rushing products to the market can sometimes be a mistake. First, markets need time to develop. Numerous examples exist where a new product is too early for the market to absorb it or where product variety has reached limits beyond which the product choice decision becomes too complicated for customers. Second, more revolutionary new product development, which often is based on significant technological advances, typically requires longer time horizons to reach completion. Putting too much emphasis on time compression may blind an organization to this basic fact. Third, the conceptual development of new product ideas requires time or "slack." In a high-speed development organization, time-compression imperatives may undermine this need. Therefore, both managers and new product developers need to find a balance between the paradoxical needs for speed and slack in their organizations. Despite its efficiency, CE will only prove to be effective when this balance is achieved through the experience and leadership of an organization's senior management.

CE can only be carried out after the integration of CAD/CAPP/CAM with the digital product models and the collaborative work environment.

Questions

1. What is the definition of concurrent engineering (CE)?
2. What is the benefit of implementing CE?
3. What is the traditional sequential engineering?
4. What are the key technologies of CE?
5. What is the Shewhart cycle?
6. What is the Six Sigma approach?
7. What is DFA&M?
8. How can a Gantt chart help you develop new products?
9. What is computer-supported collaborative work?

Tasks

1. Design a simple product as you desire. Form a group of work team. Each team member is assigned a task of design. Select one to be the leader of the team, and manage the whole product-developing process. Please record any problems raised from the CE product development, and give your solutions to deal with these problems.

References

Abarbanel B., The BOEING 777 – Concurrent engineering and digital pre-assembly, Conference proceedings of innovative applications of artificial intelligence, 1996: 1589.

Cooper R.G., Scott J.E., Portfolio Management in new product development: lessons from the leaders. Research Technology Management, 1997, 40(5): 16–52.

Debackere K., Technologies to Develop Technology. Nijmegen Innovation Lectures Monograph Series. Antwerp: Maklu Publishers, 1999.

Geoffrey B., Dewhurst P., Knight W., Product Design for Manufacture and Assembly, 2nd edition. New York: Marcel Dekker, 2002.

George S., Hout T.M., Competing against time: how time-based competition is reshaping global markets. New York: The Free Press, 1990.

http://ewh.ieee.org/soc/es/Aug1996/

http://www.referenceforbusiness.com/management/Comp-De/Concurrent-Engineering.html

Marco A., Technology Integration: making critical choices in a dynamic world. Boston: Harvard Business School Press, 1998.

ReVelle J.B., Moran J W., Charles A.C., The QFD Handbook. New York: John Wiley and Sons, 1998.

Ribbens J., Simultaneous Engineering for New Product Development: Manufacturing Applications. New York: John Wiley and Sons, 2000.

Roy U., Usher J.M., Parsaei H.R., Simultaneous Engineering: Methodologies and Applications. Amsterdam: Gordon and Breach Science Publishers, 1999.

Skalak S. C., Implementing Concurrent Engineering in Small Companies. New York: Marcel Dekker, 2002.

Ulrich K.T., Eppinger S.D., Product Design and Development, 3rd edition. Boston: McGraw-Hill/Irwin, 2004.

8 The future

Questions before you read:

1. Do you have some visions on CAD/CAPP/CAM in the future?
2. How will the CAD systems work in the future?
3. Can the CAPP systems be intelligent or not?
4. Will you still use G-code for CNC programming?
5. How do PDM and concurrent engineering help the integration of CAD/CAPP/CAM?
6. Do you know collaborative design and cloud manufacturing?

The goal of this chapter is to guide you to have an image of the future of CAD/CAPP/CAM. The next-generation 3D CAD is introduced. A knowledge-based CAD system is proposed and TRIZ logic is used as a computer-aided innovation tool. The CAPP systems for the future will have a different approach. The CAM will use standard for the exchange of product (STEP)-NC to produce manufacturing instructions. Finally, the technologies about collaborative design and cloud manufacturing are introduced as a future approach for integration of CAD/CAPP/CAM.

Imagine, for a moment, that you are working as a major CAD/CAPP/CAM developer, with massive technical resources, and more than a billion dollars in the bank. Next, also imagine that you are given a task to develop a next-generation 3D CAD/CAPP/CAM product. You don't have to generate short-term revenues and no requirement to build on the existing generation of products. Just a big toolbox full of component technologies, enough money to buy anything you need, access to an international team of smart developers, the freedom to experiment, and a reasonable amount of time to get it done. What will you create? This is more than just a fantasy exercise. You can image a new picture on a canvas for CAD/CAPP/CAM in the future.

8.1 Next-generation 3D CAD

Computer-aided design (CAD) systems have been evoluting from early two-dimensional (2D) drafting to contemporary 3D design, which are driven by both the needs for efficient design processes and high-quality representation of products, and the advancement of computing technologies and methodologies related to design practices. Many technologies in computer software have greatly impacted CAD systems, which include geometric modeling, finite element analysis, process planning, optimization algorithms, database technologies, artificial intelligence (AI), web search technologies, as well as networking and communication technologies. CAD systems are often compared with a product data management system, which is a small part of the product life-cycle management system. The paradigm behind current CAD systems can be characterized by the following features:

https://doi.org/10.1515/9783110573091-008

- Geometry, structure, and process modeling
- Graphical visualization
- Numeric analysis and simulation
- Remote collaboration through network
- Database-level functional integration
- Product life-cycle data management.

The market of commercial CAD systems is dominated by a number of large software developers who intend to offer complete solutions for the industry. Although academic research is still very active in the field of CAD systems, there are indications that the conventional system development resources will sooner or later will exhaust. A new paradigm might be necessary to provide additional support for the industry to cope with the complexity of products, processes, and data/knowledge, and to open up new opportunities for researchers, innovators, system developers, system integrators, and end users. Many recent technological developments, such as smart and ubiquitous technologies, cloud computing, semantic web, cyber-physical systems, molecular computing, quantum communications, social networking, and brain interfacing, are stimulating the discussions and research toward a new paradigm.

Approaches and solutions of next-generation CAD systems will be used by designers and engineers around and after 2025, such as novel information and knowledge-mining technologies, mobile communication and ad hoc networking, semantic network technologies, air-borne visualization technologies, smart reasoning and agent-based computing, ubiquitous sensing and computing technologies, and knowledge ontology. Natural interaction techniques will have a say in the formation of the paradigm of next-generation CAD systems.

The next-generation knowledge-based CAD systems will be cognitive, collaborative, conceptual, and creative. Cognitive refers to a specific methodology for developing CAD systems, namely, grounding the design, development, and deployment of CAD systems in cognitive studies. Collaborative indicates that design is collaborative in at least four dimensions, that is, time, space, discipline, and culture. Communication between systems, between system and human, and between humans is the core for the collaborative process. Conceptual refers to conceptual design, which mainly focuses on the understanding of the design problem and the synthesizing of design information into solution concepts. Creative represents creativity and indicates that the next-generation CAD systems will support design creativity and creative designs.

The CAD systems should be taught for the required knowledge about the design by developing a sufficient number of models as a reference resource for an effective design. This knowledge cost is one of the fundamental problems that must be solved for the next-generation CAD systems. To achieve this ambitious goal, it is necessary to conduct grounded learning of symbols by discovering patterns of functional viability based on a set of good designs. It is claimed that candidates for initial design symbols are likely the information chunks that are constituted by the inter-relations of the design variables discovered and abstracted from functionally superior designs. On the basis of the initial semantics for design symbols, the system can acquire labels by communicating with human designers.

While grounded symbols may imply a long way to the discovery of design knowledge from design experience, design rationale (DR) can be captured from design documents – a specific kind of design experience. Approaches and algorithms are designed to effectively represent and extract DR from unstructured design texts. The input of the algorithm is unstructured textual design documents, whereas the output is the DR represented by the issue, solution, and artifact level model. First, sentence relationships are modeled through language patterns by defining a sentence graph. Afterward, issue-bearing sentences, solution-bearing sentences, reason-bearing sentences, and artifact information are extracted with the help of the manifold-ranking algorithm and sentence graphs, respectively. The methods involved are text-mining, machine-learning, information retrieval, and text-processing techniques to enable automatic DR discovering.

One class of models to represent design knowledge is functional modeling, which includes function–behavior–structure, function–behavior–state, structure–behavior–function, and functional basis representations. A challenge to use those models, among others, originates from the fact that those models generally employ symbolic representations, whereas domain experts may often have to use parameter-level description to represent, design, and analyze the designed artifact.

A framework is proposed to combine both symbolic and parameter-level descriptions to represent the concept of a product in system architecting, where designers or system architects will divide, manage, and integrate large-scale design problems across engineering domains. A product development framework is proposed with the focus on hierarchical system decomposition and consistency management of design information in the conceptual design. The framework supports modeling of design knowledge about function requirements and corresponding structural and behavioral realizations based on the function–behavior–state modeling. In this framework, the symbolic design knowledge in the conceptual design is acquired from discussions among system architects and domain experts who often use both symbolic terms (natural language) and parameter-level product models. Among various kinds of design knowledge, causal knowledge plays a critical role in the evaluation of design solutions, as well as in the identification of potential problems.

The TRIZ logic can be used in developing a dialog-based computer-aided innovation system. The foundation of their approach includes the classification of problems according to their characteristics and the types of human cognitive strategies. A necessary requirement to be considered for the next-generation CAD systems is the capability to direct the thinking process of inexperienced designers in the early stages of design, that is, the formulation of design problem. In parallel to a series of research on question-based design, the dialog-based system deals with conceptual design stage. Through a dialog-based interaction, it is possible to guide the user toward a proper formulation of the problem statement, which is an essential step of any conceptual design activity. TRIZ application in particular and innovative design in general depend to a large extent on the search for potential design solutions from the existing technologies. Patent databases provide a rich source of technological information, if they can be used in an efficient and effective manner. Future CAD should provide several tools for designers to interact with patent databases, such as search functionality based on keyword, providing patent citation measures for a list of patents generated from searching, and facilitating design information extraction (functions, solutions, engineering conflicts, ontology, etc.). Designers could have multiple objectives in searching for design ideas in a patent database, such as means to satisfy specified functions, applications of particular structures or components, contradictions solved, or principles used for solving them. Regardless of the search objective and strategy employed, designers need to go through the search results to find high-value patents.

For a long time, both the academia and the industry are seeking more natural manner for designers to finish design process under a low-stress and highly effective state. A never-ending effort in CAD systems is the expression and documentation of design ideas generated during the design process.

Integrating brain computer interface into CAD systems will allow designers to naturally interact with the systems, thus allowing them to work efficiently and, at the same time, to maintain necessary flexibility for creative thinking. Most of the existing CAD systems use Windows, Icons, Mouse, and Pointer, also called WIMP, as the main user interaction tools. The designers can create and modify freeform surfaces through 3D hand gestures so that the designers need not be distracted by details of engineering design, such as edge boundary creation, surface trimming, and so on. Consequently, the designers can conduct the conceptual design following a more natural and intuitive processes.

To enable next-generation CAD tools to effectively support top-down design of products, a top-down assembly design process is refined from the traditional product design process to better exhibit the recursive-execution and structure-evolvement characteristics of product design. On the basis of the top-down assembly design process, a multilevel assembly model is put forward to capture the abstract information, skeleton information, and detailed information involved. The multilevel assembly model is a meta-level implementation and is easy to be extended. Moreover, the inheritance mechanisms are explored to ensure the feasibility of information transferring and conversion between different design phases in the top-down assembly design process. A top-down assembly design sample is analyzed at length to show the application effects of the multilevel assembly model and the relevant inheritance mechanisms.

A few issues of the next-generation CAD systems have been discussed here. We envision that the future CAD systems will hinge on two fundamental pillars: the first is designers' mental model in the design process, particularly conceptual design process, whereas the second includes technologies supporting activities underlying the entire design process. In meeting business and social requirements on the future CAD systems, the results from these fundamental explorations will enable CAD industry to implement work flows that are aligned with the designers' mental model. As such, the designers' capability to achieve better creativity and productivity may be empowered by the CAD systems with the new paradigms grounded on those fundamentals.

8.2 CAPP in another way

Because computer-aided process planning (CAPP) is such a wide area and so many technologies have been involved in research and implementation of CAPP systems, as well as the rapid development of today's computer-aided techniques, it is not easy to predict future trends.

Computer-aided manufacturing (CIM) has been a challenging concept for the manufacturing industry. CAPP has a key role to play in a CIM system. So far, interfacing CAPP with CAD and computer-aided manufacturing (CAM) systems is still a popular approach to complete the concept of CIM. A few CAPP systems have realized or are undertaking interfaces to CAD, CAM, and some other computerized systems. However, in the interfacing schemes, the present CAPP systems still have some unresolved problems that make the system unsatisfactory in application. In 3D the problems get worse because of the many incompatible ways of sorting surface and space curves. Some other attempts, such as the approach of boundary representation and the approach of constructive solid geometry tree in which the cavities are recognized from the spatial relationships between the primitive volumes, do not provide any semantic information that could be associated with the machined volumes and are based on local information. Nevertheless, great efforts have been spent on the area. The expert system rules and techniques have also been used to extract features from 3D solid model. The area of interfacing CAPP with CAD still needs to be further explored in the near future. The interfacing of CAPP with CAM and some other computerized production systems such as a numerical control (NC) path, MRP system, a production simulation system, and so on are rarely reported. But it is necessary to spend more efforts on explorations.

Intelligent manufacturing is certain to play an important role in manufacturing industry. The application of expert systems to process planning has given some promising results. In spite of the fact that the results are still very limited, they are sufficient to stimulate further research. The present, though limited, success of expert systems has proved that process planning is a proper field for AI. However, some AI techniques still need further development. Existing expert systems lack adequate mathematical calculation functions. When calculation tasks have to be performed, the expert system usually takes more time than a normal computer program. This disadvantage not only requires more computing time, but also increases the cost. There are also some other problems, such as most of the knowledge representation inference engines of the current expert systems are more system designer oriented than process planner oriented. This in a sense is the reason only a few expert process-planning systems have been utilized in real production environments. AI technologies should be further developed to improve process-planning systems. Emphasis will be given to the development of more user-friendly software products. In addition, a new generation of intelligent systems or machine learning systems will emerge. Such systems will respond to the need for continuous reteaching with the capability of monitoring actual production experiences and feeding back information to the planning system. The systems can be used for self-teaching and training for novices. This distributive planning system will tend to take over the character of the original manual planning system with intelligent software and computer systems, replacing the human skills, knowledge, and experience. The common knowledge base will tend to be segmented and individual knowledge bases will be developed at each level covering each area (factory, cell, or workstation) in manufacturing hierarchy.

As a stand-alone system, if it is portable to a personal computer (PC), the CAPP system is easily used by companies that are already equipped with PCs. Integration of the CAPP system which can be portable to a PC is very convenient and welcome for companies and research organizations in the near future. To maintain software with a widely migratory ability in the future, it is important to keep this in mind.

Despite the fact that many CAPP systems have been developed throughout the world, few of them can deal with process tolerancing and dimensioning. This might be one of the reasons that CAPPs have not shown the expected result in practice. In many cases, the dimensions used in manufacturing are not identical with the dimensions on the drawing defined by the designer for functional purposes. Therefore, the dimensions in manufacturing must be derived by appropriate calculations. In other words, it is necessary that manufacturing dimensions form chains of dimensions and tolerances compatible with drawing dimensions and tolerances, that is, a manufactured dimension must fall within the tolerance range of the corresponding design dimension. The development is still far behind the necessary requirements.

8.3 Next-generation CAM

CAM commonly refers to the use of NC computer software applications to create detailed instructions (G-code) that drive computer numerical control machine tools for manufacturing parts. With the development of STEP and STEP-NC, the products can be directly manufactured without any G-codes.

Robotic work cells are a very affordable way to cover large, multiaxis work envelopes for a wide variety of relatively low-force machining functions, such as trimming and grinding, deburring, polishing, finishing, gluing, and light cutting.

Additive and hybrid additive manufacturing techniques will revolutionize the way you think about making parts. By building up complex geometries (including internal cavities), and then machining them as needed for tight tolerances, completely new classes of parts can be manufactured.

Manufacturing is entering a dynamic new phase. By 2025, a new global consuming class will have emerged, and the majority of consumption will take place in developing economies. This will create rich new market opportunities. Meanwhile, in the established markets, demand is fragmenting as customers ask for greater variation and more types of after-sales service. A rich pipeline of innovations in materials and processes – from nano materials to 3D printing to advanced robotics – also promises to create fresh demand and drive further productivity gains across manufacturing industries and geographies.

8.4 Integrated CAD/CAPP/CAM systems in the future

Industry is on the threshold of the fourth industrial revolution, as digitalization now follows after the automation of production. The goals are to increase productivity and efficiency, speed, and quality. In this way, companies can remain competitive on the path to the future of industry.

With the development of CAD/CAPP/CAM, the virtual design and manufacturing is possible. Collaborative design and manufacturing, network-based design and manufacturing, multiagent-based design and manufacturing, and cloud manufacturing all have promising future. The infrastructure for CAD/CAPP/CAM will be more advanced. Engineers will have more choices to design and manufacture the products. The human–machine interface will change a lot. Although the CAD/CAPP/CAM systems are being considered, in the future, the function of these systems may be carried out without the aids of computer. Some products will replace computers to fulfill the collaborative work of design and manufacture. What kind of products will they be? We are waiting for those.

Questions and tasks

1. What are the latest developments in CAD/CAPP/CAM?
2. Image how the CAD/CAPP/CAM systems work in the future?

References

Alting L., Zhang H.C., Computer aided process planning: the state-of-the-art survey. International Journal of Production Research, 1989, 27(4): 553–585.

https://www.researchgate.net/publication/233748507_Fundamentals_of_next_generation_CADE_systems_Comput_Aided_Des

Index

https://doi.org/10.1515/9783110573091-009

www.ingramcontent.com/pod-product-compliance
Lightning Source LLC
Chambersburg PA
CBHW061212240926
51846CB00006B/1202